Photovoltaics: Designs, Systems and Applications

Photovoltaics: Designs, Systems and Applications

Michael Stock

Larsen & Keller
www.larsen-keller.com

Photovoltaics: Designs, Systems and Applications
Michael Stock
ISBN: 978-1-64172-089-2 (Hardback)

⊟ Larsen & Keller

Published by Larsen and Keller Education,
5 Penn Plaza,
19th Floor,
New York, NY 10001, USA

Cataloging-in-Publication Data

Photovoltaics : designs, systems and applications / Michael Stock.
    p. cm.
Includes bibliographical references and index.
ISBN 978-1-64172-089-2
1. Photovoltaic power generation. 2. Photovoltaic power systems. 3. Solar energy. I. Stock, Michael.
TK1087 .P46 2019
621.312 44--dc23

For more information regarding Larsen and Keller Education and its products, please visit the publisher's website www.larsen-keller.com

Table of Contents

Preface

Photovoltaics (PV) is the technology that converts light into electricity by using semiconducting materials that exhibit photovoltaic effect. Solar PV has the advantages of being environment friendly as it does not generate pollution or greenhouse gases. It shows simple scalability relative to power needs and the materials used for its construction are readily available. However, it requires direct sunlight or a sun-tracking system for optimal performance. A PV system consists of several components, such as solar panels, solar inverter, mounting, cabling and other electrical accessories. It can range in size and scope from small, roof-top mounted systems to large utility-scale power stations. This book explores all the important aspects of photovoltaics in the present day scenario. The various technological advancements in PV technology that have future implications are also glanced at. As this field is emerging at a rapid pace, the contents of this book will help the readers understand the modern concepts and applications of the subject.

A detailed account of the significant topics covered in this book is provided below:

Chapter 1, Photovoltaics (PV) refers to the conversion of light into electricity with the aid of semiconducting materials through the phenomenon of photovoltaic effect. The aim of this chapter is to provide an overview of photovoltaics through a detailed discussion of sun-free photovoltaics, concentrator photovoltaics, photovoltaic effect, photoelectric effect, polarizing organic photovoltaics, etc. **Chapter 2**, Among the semiconducting materials used in photovoltaics, silicon crystals are the most common. The topics elaborated in this chapter will help in developing a better perspective about the materials that are used in photovoltaics, such as crystalline silicon, polycrystalline silicon, monocrystalline silicon, cadmium telluride, etc. **Chapter 3**, Generally, a photovoltaic system is a combination of solar panels comprising of a number of solar cells. It can be ground-mounted, wall-mounted or rooftop-mounted. The mount may either be fixed or a solar tracker affixed to follow the sun across the sky. This chapter has been carefully written to provide an understanding of photovoltaic systems and their components such as photovoltaic cell, solar panel, solar shingle, solar cable, etc. **Chapter 4**, Photovoltaics is useful for a variety of applications like powering orbiting satellites and spacecraft, grid connected power generation, etc. The diverse applications of photovoltaics in the current scenario have been thoroughly discussed in this chapter, including the applications of photovoltaic appliances and transport applications.

It gives me an immense pleasure to thank our entire team for their efforts. Finally in the end, I would like to thank my family and colleagues who have been a great source of inspiration and support.

Michael Stock

Introduction to Photovoltaics

Photovoltaics (PV) refers to the conversion of light into electricity with the aid of semiconducting materials through the phenomenon of photovoltaic effect. The aim of this chapter is to provide an overview of photovoltaics through a detailed discussion of sun-free photovoltaics, concentrator photovoltaics, photovoltaic effect, photoelectric effect, polarizing organic photovoltaics, etc.

The word Photovoltaic is a combination of the Greek word for Light and the name of the physicist Alessandro Volta. It identifies the direct conversion of sunlight into energy by means of solar cells. The conversion process is based on the photoelectric effect discovered by Alexander Becquerel in 1839. The photoelectric effect describes the release of positive and negative charge carriers in a solid state when light strikes its surface.

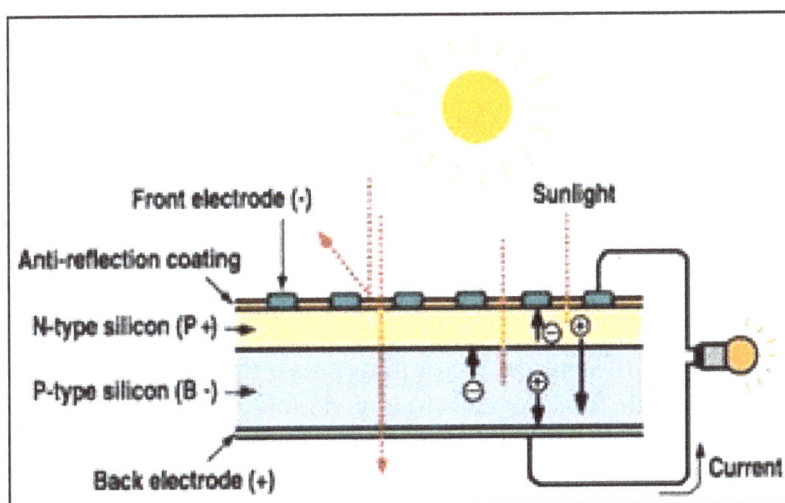

To create the PV effect, radiation from the sun ('sunlight') hits a photovoltaic cell. These cells are made up of two layers of semi-conducting material, typically silicon, that have been chemically treated. The industry refers to these layers as P and N. The boundary between P and N acts as a diode allowing electrons to move from N to P, but not from P to N. When photons with sufficient energy hit the cell, they cause electrons to move (from N to P only) causing excess electrons in the N-layer and a shortage in the P layer.

This voltage difference is typically in the range of 0.5V for as long as the cell is in sunlight. If you short-circuit the upper and lower layer, a current runs of about 3 Amps. If you arrange sufficient cells in series, the result is a PV module or PV panel. Let's say 36 cells in series produce 36 x 0.5V = 18V at 3 Amps = 54Watts.

The following graphic sets out the layers within the cell. The top layer is an Anti-Reflective-Coating (ARC) that enhances the light effect of the sun. The N layer is typically semi-conducting silicon doped with phosphorus that creates the free flow of electrons. The P layer is again typically semi-conducting silicon, but this time doped with boron which creates the free flow of positive charges called "holes". As the holes and electrons are attracted and move towards each other, they create an electrical field across the P-N junction. Sunlight striking this electrical field separates the electrons and holes, creating the voltage.

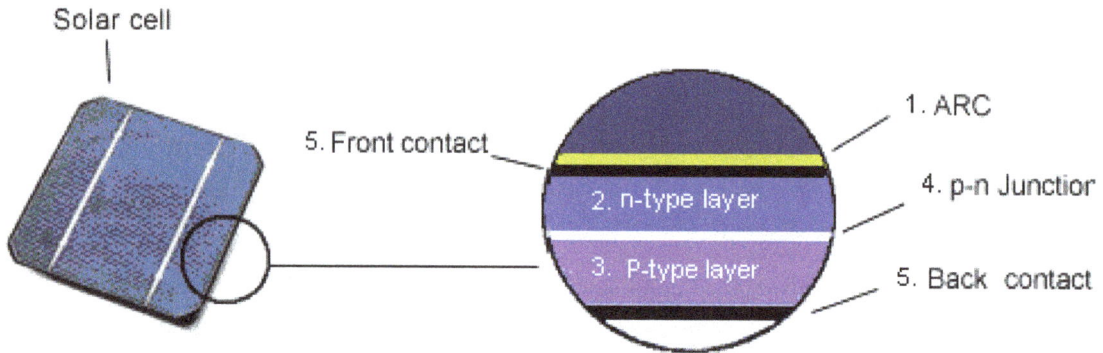

The voltage pushes the flow of electrons or 'DC current' to contacts at the front and back of the cell where it is conducted away along the wiring circuitry that connects the cells together.

Solar Cells and Arrays

Solar cells are typically combined into modules that hold about 40 cells; about 10 of these modules are mounted in PV arrays that can measure up to several meters on a side. These flat-plate PV arrays can be mounted at a fixed angle facing south, or they can be mounted on a tracking device that follows the sun, allowing them to capture the most sunlight over the course of a day. About 10 to 20 PV arrays can provide enough power for a household; for large electric utility or industrial applications, hundreds of arrays can be interconnected to form a single, large PV system.

Thin film solar cells use layers of semiconductor materials only a few micrometers thick. Thin film technology has made it possible for solar cells to now double as rooftop shingles, roof tiles, building facades, or the glazing for skylights or atria. The solar cell version of items such as shingles offer the same protection and durability as ordinary asphalt shingles.

Some solar cells are designed to operate with concentrated sunlight. These cells are built into concentrating collectors that use a lens to focus the sunlight onto the cells. This approach has both advantages and disadvantages compared with flat-plate PV arrays. The main idea is to use very little of the expensive semiconducting PV material while collecting as much sunlight as possible. But because the lenses must be pointed at the sun, the use of concentrating collectors is limited to the sunniest parts of the country.

Efficiency

The performance of a solar cell is measured in terms of its efficiency at turning sunlight into electricity. Only sunlight of certain energies will work efficiently to create electricity, and much of it is reflected or absorbed by the material that make up the cell. Because of this, a typical commercial solar cell has an efficiency of 15% about one-sixth of the sunlight striking the cell generates electricity, although leading competitors are working towards 18%. The theoretical maximum efficiency of a solar cell using current techniques is in the 30% range.

For this reason, implementations of photovoltaics are most effective in areas with a large amount of daily sunlight. Note that the solar cells are temperature-dependent, such that in a cold environment a photovoltaic cell performs better than in a hot environment. (0.3% increase per 1 degree C drop in temperature). Unfortunately, there aren't a lot of places on earth that are both cold and have long sunny days.

Low efficiencies mean that larger arrays are needed, and that means higher cost. Improving solar cell efficiencies while holding down the cost per cell is an important goal of the all participants in the solar energy industry, and they have made significant progress. The first solar cells, built in the 1950s, had efficiencies of less than 4%.

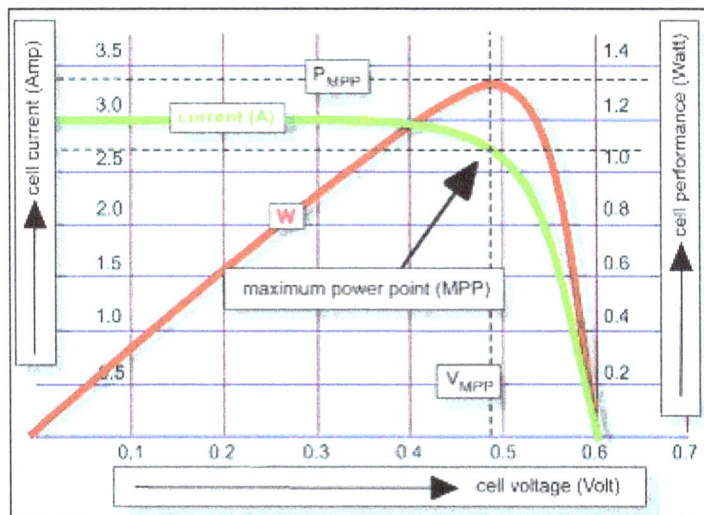

The diagram is sets out the maximum power performance of a photovoltaic cell. The red curve is the voltage-performance graph of the cell and the green curve is the current-voltage graph. The best performance (in Watts) is obtained at that voltage at which the current definitely starts to decline: the maximum power point (MPP).

PV Materials–Silicon and Thin Film Technologies

Three key elements in a solar cell form the basis of their manufacturing technology. The first is the semiconductor, which absorbs light and converts it into electron-hole pairs. The second is the semiconductor junction, which separates the photo-generated carriers (electrons and holes), and the third is the contacts on the front and back of the cell that allow the current to flow to the external circuit. The two main categories of technology are defined by the choice of the semiconductor: either crystalline silicon in a wafer form or thin films of other materials.

Crystalline silicon cells: Crystalline silicon solar cells currently represent over 90% of the world market for solar cells. Historically, crystalline silicon (c-Si) has been used as the light-absorbing semiconductor in most solar cells, even though it is a relatively poor absorber of light and requires a considerable thickness (several hundred microns) of material. Nevertheless, it has proved convenient because it yields stable solar cells with good efficiencies (11-16%, half to two-thirds of the theoretical maximum) and uses process technology developed from the huge knowledge base of the microelectronics industry.

Two types of crystalline silicon are used in the industry. The first is monocrystalline, produced by slicing wafers (up to 150mm diameter and 350 microns thick) from a high-purity single crystal boule. The second is multicrystalline silicon, made by sawing a cast block of silicon first into bars and then wafers. The main trend in crystalline silicon cell manufacture is toward multicrystalline technology. Monocrystalline cells have higher conversion efficiency than multicrystalline cells, but costs of monocrystalline wafers are generally higher than multicrystalline wafers.

For both mono- and multicrystalline Si, a semiconductor homojunction is formed by diffusing phosphorus (an n-type dopant) into the top surface of the boron doped (p-type) Si wafer. Screen-printed contacts are applied to the front and rear of the cell, with the front contact pattern specially designed to allow maximum light exposure of the Si material with minimum electrical (resistive) losses in the cell.

The most efficient production cells use monocrystalline c-Si with laser grooved, buried grid contacts for maximum light absorption and current collection.

Some companies are using technologies that by-pass some of the inefficiencies of the crystal growth/casting and wafer sawing route. One route is to grow a ribbon of silicon, either as a plain two-dimensional strip or as an octagonal column, by pulling it from a silicon melt.

Another is to melt silicon powder on a cheap conducting substrate. These processes may bring with them other issues of lower growth/pulling rates and poorer uniformity and surface roughness.

Each c-Si cell generates about 0.5V. So, 36 cells are usually soldered together in series to produce a module with an output to charge a 12V battery. The cells are hermetically sealed under toughened, high transmission glass to produce highly reliable, weather resistant modules.

Thin film solar cells represent a small, but fast growing segment of the PV market. The high cost of crystalline silicon wafers (they make up 40-50% of the cost of a finished module) has led the industry to look at cheaper materials to make solar cells.

The selected materials are all strong light absorbers and only need to be about 1micron thick, so materials costs are significantly reduced. The most common materials are amorphous silicon (a-Si, still silicon, but in a different form), or the polycrystalline materials: cadmium telluride (CdTe) and copper indium (gallium) diselenide (CIS or CIGS).

Each of these three is amenable to large area deposition (on to substrates of about 1 meter dimensions) and hence high volume manufacturing. The thin film semiconductor layers are deposited on to either coated glass or stainless steel sheet.

The semiconductor junctions are formed in different ways, either as a p-i-n device in amorphous silicon, or as a hetero-junction (e.g. with a thin cadmium sulphide layer) for CdTe and CIS. A transparent conducting oxide layer (such as tin oxide) forms the front electrical contact of the cell, and a metal layer forms the rear contact.

Thin film technologies are all complex. They have taken at least twenty years, supported in some cases by major corporations, to get from the stage of promising research (about 8% efficiency at 1cm^2 scale) to the first manufacturing plants producing early product.

Amorphous silicon is the well-developed of the thin film technologies. In its simplest form, the cell structure has a single sequence of p-i-n layers. Such cells suffer from significant degradation in their power output (in the range 15-35%) when exposed to the sun.

The mechanism of degradation is called the Staebler-Wronski Effect, after its discoverers. Better stability requires the use of a thinner layer in order to increase the electric field strength across the material. However, this reduces light absorption and hence cell efficiency.

This has led the industry to develop tandem and even triple layer devices that contain p-i-n cells stacked one on top of the other. In the cell at the base of the structure, the a-Si is sometimes alloyed with germanium to reduce its band gap and further improve light absorption. All this added complexity has a downside though; the processes are more complex and process yields are likely to be lower.

In order to build up a practically useful voltage from thin film cells, their manufacture usually includes a laser scribing sequence that enables the front and back of adjacent cells to be directly interconnected in series, with no need for further solder connection between cells.

As before, thin film cells are laminated to produce a weather resistant and environmentally robust module. Although they are less efficient (production modules range from 5 to 8%), thin films are potentially cheaper than c-Si because of their lower materials costs and larger substrate size.

However, some thin film materials have shown degradation of performance over time and stabilized efficiencies can be 15-35% lower than initial values. Many thin film technologies have demonstrated best cell efficiencies at research scale above 13%, and best prototype module efficiencies above 10%. The technology that is most successful in achieving low manufacturing costs in the long run is likely to be the one that can deliver the highest stable efficiencies (probably at least 10%) with the highest process yields.

Amorphous silicon is the most well-developed thin film technology to-date and has an interesting avenue of further development through the use of "microcrystalline" silicon which seeks to combine the stable high efficiencies of crystalline Si technology with the simpler and cheaper large area deposition technology of amorphous silicon.

However, conventional c-Si manufacturing technology has continued its steady improvement year by year and its production costs are still falling too.

Advantages

- Electricity produced by solar cells is clean and silent. Because they do not use fuel other than sunshine, PV systems do not release any harmful air or water pollution into the environment, deplete natural resources, or endanger animal or human health.

- Photovoltaic systems are quiet and visually unobtrusive.

- Small-scale solar plants can take advantage of unused space on rooftops of existing buildings.

- PV cells were originally developed for use in space, where repair is extremely expensive, if not impossible. PV still powers nearly every satellite circling the earth because it operates reliably for long periods of time with virtually no maintenance.

- Solar energy is a locally available renewable resource. It does not need to be imported from other regions of the country or across the world. This reduces environmental impacts associated with transportation and also reduces our dependence on imported oil. And, unlike fuels that are mined and harvested, when we use solar energy to produce electricity we do not deplete or alter the resource.

- A PV system can be constructed to any size based on energy requirements. Furthermore, the owner of a PV system can enlarge or move it if his or her energy needs change. For instance, homeowners can add modules every few years as their energy usage and financial resources grow. Ranchers can use mobile trailer-mounted pumping systems to water cattle as the cattle are rotated to different fields.

Disadvantages

- Some toxic chemicals, like cadmium and arsenic, are used in the PV production process. These environmental impacts are minor and can be easily controlled through recycling and proper disposal.

- Solar energy is somewhat more expensive to produce than conventional sources of energy due in part to the cost of manufacturing PV devices and in part to the conversion efficiencies of the equipment. As the conversion efficiencies continue to increase and the manufacturing costs continue to come down, PV will become increasingly cost competitive with conventional fuels.

- Solar power is a variable energy source, with energy production dependent on the sun. Solar facilities may produce no power at all some of the time, which could lead to an energy shortage if too much of a region's power comes from solar power.

Sun-free Photovoltaics

A new photovoltaic energy-conversion system developed at MIT can be powered solely by heat, generating electricity with no sunlight at all. While the principle involved is not new, a novel way of engineering the surface of a material to convert heat into precisely tuned wavelengths of light — selected to match the wavelengths that photovoltaic cells can best convert to electricity — makes the new system much more efficient than previous versions

The key to this fine-tuned light emission, lies in a material with billions of nanoscale pits etched on its surface. When the material absorbs heat — whether from the sun, a hydrocarbon fuel, a decaying radioisotope or any other source — the pitted surface radiates energy primarily at these carefully chosen wavelengths.

Based on that technology, MIT researchers have made a button-sized power generator fueled by butane that can run three times longer than a lithium-ion battery of the same weight; the device can then be recharged instantly, just by snapping in a tiny cartridge of fresh fuel. Another device, powered by a radioisotope that steadily produces heat from radioactive decay, could generate electricity for 30 years without refueling or servicing — an ideal source of electricity for spacecraft headed on long missions away from the sun.

According to the U.S. Energy Information Administration, 92 percent of all the energy we use involves converting heat into mechanical energy, and then often into electricity — such as using fuel to boil water to turn a turbine, which is attached to a generator. But today's mechanical systems have relatively low efficiency, and can't be scaled down to the small sizes needed for devices such as sensors, smartphones or medical monitors.

Being able to convert heat from various sources into electricity without moving parts would bring huge benefits, especially if we could do it efficiently, relatively inexpensively and on a small scale.

It has long been known that photovoltaic (PV) cells needn't always run on sunlight. Half a century ago, researchers developed Thermo photovoltaics (TPV), which couple a PV cell with any source of heat: A burning hydrocarbon, for example, heats up a material called the thermal emitter, which radiates heat and light onto the PV diode, generating electricity. The thermal emitter's radiation includes far more infrared wavelengths than occur in the solar spectrum, and "low band-gap" PV materials invented less than a decade ago can absorb more of that infrared radiation than standard silicon PVs can. But much of the heat is still wasted, so efficiencies remain relatively low.

An Ideal Match

The solution, Celanovic says, is to design a thermal emitter that radiates only the wavelengths that the PV diode can absorb and convert into electricity, while suppressing other wavelengths. "But how do we find a material that has this magical property of emitting only at the wavelengths that we want?" The answer: Make a photonic crystal by taking a sample of material and create some nanoscale features on its surface — say, a regularly repeating pattern of holes or ridges — so light propagates through the sample in a dramatically different way.

"By choosing how we design the nanostructure, we can create materials that have novel optical properties," Soljačić says. This gives us the ability to control and manipulate the behavior of light.

The team — which also includes Peter Bermel, research scientist in the Research Laboratory for Electronics (RLE); Peter Fisher, professor of physics; and Michael Ghebrebrhan, a postdoc in RLE — used a slab of tungsten, engineering billions of tiny pits on its surface. When the slab heats up, it generates bright light with an altered emission spectrum because each pit acts as a resonator, capable of giving off radiation at only certain wavelengths.

This powerful approach — co-developed by John D. Joannopoulos, the Francis Wright Davis Professor of Physics and ISN director, and others — has been widely used to improve lasers, light-emitting diodes and even optical fibers. The MIT team, supported in part by a seed grant from the MIT Energy Initiative, is now working with collaborators at MIT and elsewhere to use it to create several novel electricity-generating devices.

This approach to producing miniature power supplies could lead to lighter portable electronics, which is critical for the soldier to lighten his load. It not only reduces his burden, but also reduces the logistics chain to deliver those devices to the field. There are a lot of lives at stake, so if you can make the power sources more efficient, it could be a great benefit.

The button-like device that uses hydrocarbon fuels such as butane or propane as its heat source — known as a micro-TPV power generator — has at its heart a "micro-reactor" designed by Klavs Jensen, While the device achieves a fuel-to-electricity conversion efficiency three times greater than that of a lithium-ion battery of the same size and weight, At that point, our TPV generator could power your smartphone for a whole week without being recharged.

Concentrator Photovoltaics

One of the ways to increase the output from the photovoltaic systems is to supply concentrated light onto the PV cells. This can be done by using optical light collectors, such as lenses or mirrors. The PV systems that use concentrated light are called concentrating photovoltaics (CVP). The CPV collect light from a larger area and concentrate it to a smaller area solar cell. This is illustrated in figure.

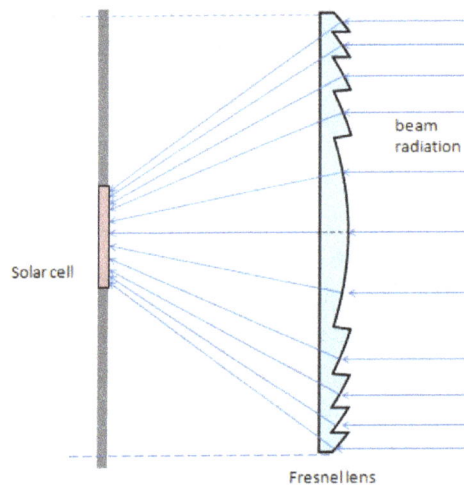

Figure: This is one of the common types of concentrator cells based on Fresnel lens, which takes the parallel beam sunlight and directs it to a small area. For an effective use of the Fresnel CPV system, two-axis sun tracking is needed to ensure that the rays are perpendicular to the lens.

Lower efficiency CPV technologies may employ silicon, CdTe, and CIGS (copper indium gallium selenide) cells, but the highest efficiencies can be achieved with multi-junction cells. Field efficiencies for these multi-junction cells are in the 30% range, and laboratory tests have achieved upwards of 40% efficiency.

The CPV can only use direct beam radiation and cannot use diffuse radiation (diffused from clouds and atmosphere). Therefore, these systems are suited best for areas with high direct normal irradiance. For proper light concentration, sun tracking is needed for achieving high cell performance. Tracking is especially critical for high concentration systems. In general, the CPV can be classified into low-concentration, medium-concentration, and high-concentration.

Table: Different classes of CPV systems and their requirements

	Low-concentration	Medium-concentration	High-concentration
Concentration ratio	2-10	10-100	100-400(and above)

PV materials	Silicon	Silicon, CdTe, etc.	Multi junction cells
Cooling	not required	Passive cooling	Active cooling
Tracking	not required	1-axis tracking	2-axis tracking

The high concentration of sunlight achieved with multi junction cells requires more sophisticated cooling and tracking systems, which can potentially result in higher energy costs.

CPV technology is expected to grow and to expand on market. The cost effectiveness of CPV technology is related to the fact that much smaller sized solar cells are used to convert the concentrated light, which means that much less expensive PV semiconductor material is used. Also, the optics added to the system are made from glass and are usually less expensive than the cells themselves.

Figure: Photovoltaic power system on the roof of the St. Petersburg Academic University - Nanotechnology Centre of RAS. In the center, there is a typical design of the Fresnel CPV system represented by a module of multiple cells, with a separate Fresnel lens placed on top of each single cell. On the right side of the image, a regular PV silicon cell module is shown.

Challenges

Modern CPV systems operate most efficiently in highly concentrated sunlight (i.e. concentration levels equivalent to hundreds of suns), as long as the solar cell is kept cool through the use of heat sinks. Diffuse light, which occurs in cloudy and overcast conditions, cannot be highly concentrated using conventional optical components only (i.e. macroscopic lenses and mirrors). Filtered light, which occurs in hazy or polluted conditions, has spectral variations which produce mismatches between the electrical currents generated within the series-connected junctions of spectrally "tuned" multi-junction (MJ) photovoltaic cells. These CPV features lead to rapid decreases in power output when atmospheric conditions are less than ideal.

To produce equal or greater energy per rated watt than conventional PV systems, CPV systems must be located in areas that receive plentiful direct sunlight. This is typically specified as average DNI greater than 5.5-6 kWh/m²/day or 2000kWh/m²/yr. Otherwise, evaluations of annualized DNI vs. GNI/GHI irradiance data have concluded that conventional PV should still perform better over time than presently available CPV technology in most regions of the world.

CPV Strengths	CPV Weaknesses
High efficiencies under direct normal irradiance	HCPV cannot utilize diffuse radiation. LCPV can only utilize a fraction of diffuse radiation.
Low cost per watt of manufacturing capital	Power output of MJ solar cells is more sensitive to shifts in radiation spectra caused by changing atmospheric conditions.
Low temperature coefficients	Tracking with sufficient accuracy and reliability is required.
No cooling water required for passively cooled systems	May require frequent cleaning to mitigate soiling losses, depending on the site
Additional use of waste heat possible for systems with active cooling possible (e.g.large mirror systems)	Limited market – can only be used in regions with high DNI, cannot be easily installed on rooftops
Modular – kW to GW scale	Strong cost decrease of competing technologies for electricity production
Increased and stable energy production throughout the day due to (two-axis) tracking	Bankability and perception issues
Low energy payback time	New generation technologies, without a history of production (thus increased risk)
Potential double use of land e.g. for agriculture, low environmental impact	Optical losses
High potential for cost reduction	Lack of technology standardization
Opportunities for local manufacturing	–
Smaller cell sizes could prevent large fluctuations in module price due to variations in semiconductor prices	–
Greater potential for efficiency increase in the future compared to single-junction flat plate systems could lead to greater improvements in land area use, BOS costs, and BOP costs	–

Ongoing Research and Development

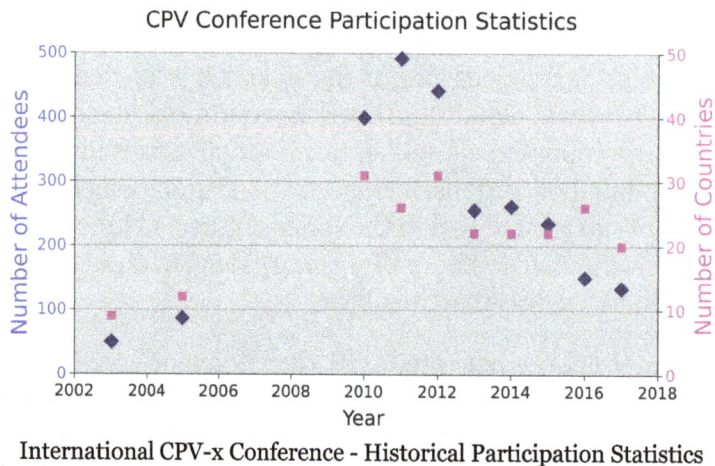

International CPV-x Conference - Historical Participation Statistics

CPV research and development has been pursued in over 20 countries for more than a decade. The annual CPV-x conference series has served as a primary networking and exchange forum between university, government lab, and industry participants. Government agencies have also continued to encourage a number of specific technology thrusts.

ARPA-E announced a first round of R&D funding in late 2015 for the MOSAIC Program (Microscale Optimized Solar-cell Arrays with Integrated Concentration) to further combat the location and expense challenges of existing CPV technology. As stated in the program description: "MOSAIC projects are grouped into three categories: complete systems that cost effectively integrate micro-CPV for regions such as sunny areas of the U.S. southwest that have high Direct Normal Incident (DNI) solar radiation; complete systems that apply to regions, such as areas of the U.S. Northeast and Midwest, that have low DNI solar radiation or high diffuse solar radiation; and concepts that seek partial solutions to technology challenges."

In Europe, the CPVMATCH Program (Concentrating PhotoVoltaic Modules using Advanced Technologies and Cells for Highest efficiencies) aims "to bring practical performance of HCPV modules closer to theoretical limits". Efficiency goals achievable by 2019 are identified as 48% for cells and 40% for modules at >800x concentration.

The Australian Renewable Energy Agency (ARENA) extended its support in 2017 for further commercialization of the HCPV technology developed by Raygen. Their 250kW dense array receivers are the most powerful CPV receivers thus far created, with demonstrated PV efficiency of 40.4% and include usable heat co-generation.

Optical Design

The design of macroscopic sunlight concentrators for CPV introduces a very specific optical design problem, with features that makes it different from any other optical design. It has to be efficient, suitable for mass production, capable of high concentration, insensitive to manufacturing and mounting inaccuracies, and capable of providing uniform illumination of the cell. All these reasons make nonimaging optics the most suitable for CPV.

For very low concentrations, the wide acceptance angles of nonimaging optics avoid the need for active solar tracking. For medium and high concentrations, a wide acceptance angle can be seen as a measure of how tolerant the optic is to imperfections in the whole system. It is vital to start with a wide acceptance angle since it must be able to accommodate tracking errors, movements of the system due to wind, imperfectly manufactured optics, imperfectly assembled components, finite stiffness of the supporting structure or its deformation due to aging, among other factors. All of these reduce the initial acceptance angle and, after they are all factored in, the system must still be able to capture the finite angular aperture of sunlight.

Types

CPV systems are categorized according to the amount of their solar concentration, measured in "suns" (the square of the magnification).

Low Concentration PV (LCPV)

Low concentration PV are systems with a solar concentration of 2–100 suns. For economic reasons, conventional or modified silicon solar cells are typically used, and, at these concentrations, the heat flux is low enough that the cells do not need to be actively cooled. There is now modeling and experimental evidence that standard solar modules do not need any modification, tracking or cooling if the concentration level is low and yet still have increased output of 35% or more.

An example of a Low Concentration PV Cell's surface, showing the glass lensing

Medium Concentration PV

From concentrations of 100 to 300 suns, the CPV systems require two-axis solar tracking and cooling (whether passive or active), which makes them more complex.

A 10×10 mm HCPV solar cell

High Concentration Photovoltaics (HCPV)

High concentration photovoltaics (HCPV) systems employ concentrating optics consisting of dish reflectors or fresnel lenses that concentrate sunlight to intensities of 1,000 suns or more. The solar cells require high-capacity heat sinks to prevent thermal destruction and to manage temperature related electrical performance and life expectancy losses. To further exacerbate the concentrated cooling design, the heat sink must be passive, otherwise the power required for active cooling will reduce the overall conversion efficiency and economy. Multi-junction solar cells are currently favored over single junction cells, as they are more efficient and have a lower temperature coefficient (less loss in efficiency with an increase in temperature). The efficiency of both cell types rises with increased concentration; multi-junction efficiency rises faster. Multi-junction solar cells, originally designed for non-concentrating PV on space-based satellites, have been re-designed due to the high-current density encountered with CPV (typically 8 A/cm² at 500 suns). Though the cost of multi-junction solar cells is roughly 100 times that of conventional silicon cells of the same area, the small cell area employed makes the relative costs of cells in each system comparable and the system economics favor the multi-junction cells. Multi-junction cell efficiency has now reached 44% in production cells.

The 44% value given above is for a specific set of conditions known as "standard test conditions". These include a specific spectrum, an incident optical power of 850 W/m², and a cell temperature of 25°C. In a concentrating system, the cell will typically operate under conditions of variable spectrum, lower optical power, and higher temperature. The optics needed to concentrate the light have limited efficiency themselves, in the range of 75–90%. Taking these factors into account, a solar module incorporating a 44% multi-junction cell might deliver a DC efficiency around 36%. Under similar conditions, a crystalline silicon module would deliver an efficiency of less than 18%.

When high concentration is needed (500–1000 times), as occurs in the case of high efficiency multi-junction solar cells, it is likely that it will be crucial for commercial success at the system level to achieve such concentration with a sufficient acceptance angle. This allows tolerance in mass production of all components, relaxes the module assembling and system installation, and decreasing the cost of structural elements. Since the main goal of CPV is to make solar energy inexpensive, there can be used only a few surfaces. Decreasing the number of elements and achieving high acceptance angle, can be relaxed optical and mechanical requirements, such as accuracy of the optical surfaces profiles, the module assembling, the installation, the supporting structure, etc. To this end, improvements in sunshape modelling at the system design stage may lead to higher system efficiencies.

Installations

Concentrator photovoltaics technology has established its presence in the solar industry over the last several years. The first CPV power plant that exceeded 1 MW-level was commissioned in Spain in 2006. By the end of 2015, the number of CPV power plants around the world accounted for a total installed capacity of 350 MW. Field data collected over the past six years is also starting to benchmark the prospects for long-term system reliability.

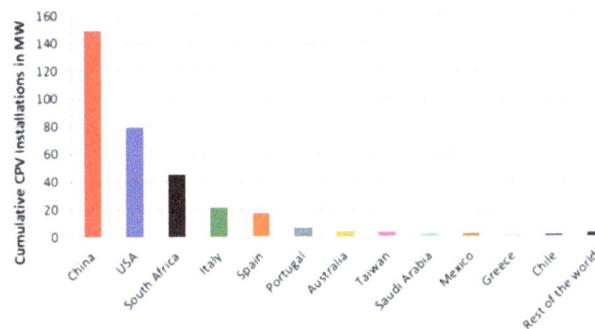

Cumulative CPV Installations in MW by country by November 2014

The emerging CPV segment has comprised ~0.1% of the fast-growing utility market for PV installations over the past decade. Unfortunately, by the end of 2015, the near term outlook for CPV industry growth has faded with closure of all of the largest CPV manufacturing facilities: including those of Suncore, Soitec, Amonix, and Solfocus. Nevertheless, the growth outlook for the overall PV industry continues to appear strong.

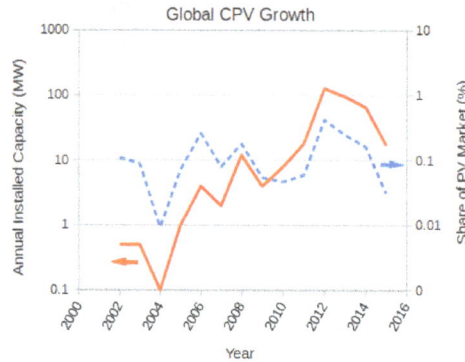

Yearly Installed CPV Capacity in MW from 2002 to 2015.

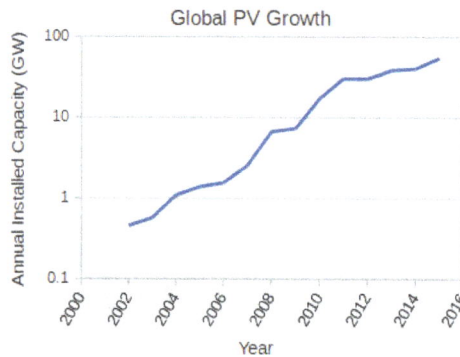

Yearly Installed PV Capacity in GW from 2002 to 2015.

List of large CPV systems

The largest CPV power plant currently in operation is of 80 MW_p capacity located in Golmud, China, hosted by Suncore Photovoltaics.

Power station	Capacity (MW_p)	Location	Vendor/Builder
Golmud 2	79.83	Golmud/Qinghai province/China	Suncore
Golmud 1	57.96	Golmud/Qinghai province/China	Suncore
Touwsrivier	44.19	Touwsrivier/Western Cape/South Africa	Soitec
Alamosa Solar Project	35.28	Alamosa, Colorado/San Luis Valley/United States	Amonix

Concentrated Photovoltaics and Thermal

Concentrator photovoltaics and thermal (CPVT), also sometimes called combined heat and power solar (CHAPS) or hybrid thermal CPV, is a cogeneration or micro cogeneration technology used in the field of concentrator photovoltaics that produces usable heat and electricity within the same system. CPVT at high concentrations of over 100 suns (HCPVT) utilizes similar components as HCPV, including dual-axis tracking and multi-junction photovoltaic cells. A fluid actively cools the integrated thermal–photovoltaic receiver, and simultaneously transports the collected heat.

Typically, one or more receivers and a heat exchanger operate within a closed thermal loop. To maintain efficient overall operation and avoid damage from thermal runaway, the demand for heat from the secondary side of the exchanger must be consistently high. Under such optimal operating conditions, collection efficiencies exceeding 70% (up to ~35% electric, ~40% thermal for HCPVT) are anticipated. Net operating efficiencies may be substantially lower depending on how well a system is engineered to match the demands of the particular thermal application.

The maximum temperature of CPVT systems is typically too low alone to power a boiler for additional steam-based cogeneration of electricity. Such systems may be economical to power lower temperature applications having a constant high heat demand. The heat may be employed in district heating, water heating and air conditioning, desalination or process heat. For applications having lower or intermittent heat demand, a system may be augmented with a switchable heat dump to the external environment in order to maintain reliable electrical output and safeguard cell life, despite the resulting reduction in net operating efficiency.

HCPVT active cooling enables the use of much higher power thermal–photovoltaic receiver units, generating typically 1–100 kilowatts electric, as compared to HCPV systems that mostly rely upon passive cooling of single ~20W cells. Such high-power receivers utilize dense arrays of cells mounted on a high-efficiency heat sink. Minimizing the number of individual receiver units is a simplification that should ultimately yield improvement in the overall balance of system costs, manufacturability, maintainability/upgradeability, and reliability.

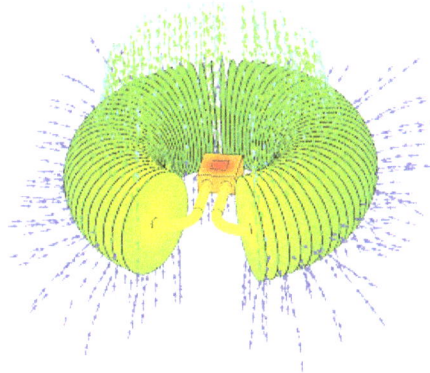

This 240 x 80 mm 1,000 suns CPV heat sink design thermal animation, was created using high resolution CFD analysis, and shows temperature contoured heat sink surface and flow trajectories as predicted.

Reliability Requirements

The maximum operating temperatures ($T_{max\ cell}$) of CPVT systems are limited to less than approximately 100–125°C on account of the intrinsic reliability limitation of their multi-junction PV cells. This contrasts to CSP and other CHP systems which may be designed to function at temperatures in excess of several hundred degrees. More specifically, the multi-junction photovoltaic cells are fabricated from a layering of thin-film III-V semiconductor materials having intrinsic lifetimes during CPV operation that rapidly decrease with an Arrhenius-type temperature dependence. The system receiver must therefore provide for highly efficient and uniform cell cooling, where an ideal receiver would provide $T_{max\ coolant} \sim T_{max\ cell}$. In addition to material and design limitations in receiver heat-transfer performance, numerous extrinsic factors, such as the frequent system thermal cycling, further reduce the practical $T_{max\ coolant}$ compatible with long system life to below about 80°C.

The higher capital costs, lesser standardization, and added engineering & operational complexities (in comparison to zero and low-concentration PV technologies) make demonstrations of system reliability and long-life performance critical challenges for the first generation of CPV and CPVT technologies. Performance certification testing standards (e.g. IEC 62108, UL 8703, IEC 62789, IEC 62670) include stress conditions that may be useful to uncover some predominantly infant and early life (<1–2 year) failure modes at the system, module, and sub-component levels. However, such standardized tests – as typically performed on only a small sampling of units – are generally incapable to evaluate comprehensive long-term (10 to 25 or more years) lifetimes for each unique CPVT system design and application under its broader range of actual operating conditions. Long-life performance of these complex systems is therefore assessed in the field, and is improved through aggressive product development cycles which are guided by the results of accelerated component/system aging, enhanced performance monitoring diagnostics, and failure analysis. Significant growth in the deployment of CPV and CPVT can be anticipated once the long-term performance and reliability concerns are better addressed to build confidence in system bankability.

Demonstration Projects

The economics of a mature CPVT industry is anticipated to be competitive, despite the large recent cost reductions and gradual efficiency improvements for conventional silicon PV (which can be installed alongside conventional CSP to provide for similar electrical+thermal generation capabilities). CPVT may currently be economical for niche markets having all of the following application characteristics:

- High solar direct normal incidence (DNI),

- Tight space constraints for placement of a solar collector array,

- High and constant demand for low-temperature (<80°C) heat,

- High cost of grid electricity,

- Access to backup sources of power or cost-efficient storage (electrical and thermal).

Utilization of a power purchase agreement (PPA), government assistance programs, and innovative financing schemes are also helping potential manufacturers and users to mitigate the risks of early CPVT technology adoption.

CPVT equipment offerings ranging from low (LCPVT) to high (HCPVT) concentration are now being deployed by several startup ventures. As such, longer-term viability of the technical and/or business approach being pursued by any individual system provider is typically speculative. Notably, the minimum viable products of startups can vary widely in their attention to reliability engineering. Nevertheless, the following incomplete compilation is offered to assist with the identification of some early industry trends.

LCPVT systems at ~14x concentration using reflective trough concentrators, and receiver pipes clad with silicon cells having dense interconnects, have been assembled by Cogenra with a claimed

75% efficiency (~15-20% electric, 60% thermal). Several such systems are in operation for more than 5 years as of 2015, and similar systems are being produced by Absolicon and Idhelio at 10x and 50x concentration, respectively.

HCPVT offerings at over 700x concentration have more recently emerged, and may be classified into three power tiers. Third tier systems are distributed generators consisting of large arrays of ~20W single-cell receiver/collector units, similar to those previously pioneered by Amonix and SolFocus for HCPV. Second tier systems utilize localized dense-arrays of cells that produce 1-100 kW of electrical power output per receiver/generator unit. First tier systems exceed 100 kW of electrical output and are most aggressive in targeting the utility market.

Several HCPVT system providers are listed in the following table. Nearly all are early demonstration systems which have been in service for under 5 years as of 2015. Collected thermal power is typically 1.5x-2x the rated electrical power.

Provider	Country	Concentrator Type	Unit Size in kW$_e$	
			Generator	Receiver
		- Tier 1 -		
Raygen	Australia	Large Heliostat Array	250	250
		- Tier 2 -		
Zenith Solar/Suncore	Israel/China/USA	Large Dish	4.5	2.25
Sun Oyster	Germany	Large Trough + Lens	4.7	2.35
Rehnu	United States	Large Dish	6.4	0.8
Airlight Energy/dsolar	Switzerland	Large Dish	12	12
Solartron	Canada	Large Dish	20	20
Southwest Solar	United States	Large Dish	20	20
		- Tier 3 -		
Silex Power	Malta	Small Dish Array	16	0.04
Solergy	Italy/USA	Small Lens Array	20	0.02

Benefits of CPV Systems

CPV offers a number of benefits to utilities:

CPV is perfectly suited for high ambient temperatures. Due to the very low temperature coefficient of the III-V multi-junction concentrator solar cell, the performance of CPV systems is much less affected by temperature than any other PV technology, i.e., the loss of efficiency is approximately one third that of crystalline silicon modules. This attribute is extremely important for the best solar sites in the world, which are generally also located in hot climates. Because of the low temperature coefficient, the efficiency and the electricity production of CPV systems are only slightly affected by high ambient temperatures in comparison to other PV technologies. The efficiency drop for a temperature difference of 40K (e.g., from 25°C at standard testing conditions to typical operating cell temperatures of 65°C) is by far the smallest for CPV systems.

CPV provides the power at the right time. CPV systems using a high concentration are always tracked using two axes. Two-axis tracking allows for a homogeneous electricity production profile over the day because the panels are always oriented perpendicularly to the incident irradiation from the sun. The most important effect is that the power production is at high levels when the power demand peaks in the afternoon. Afternoon peaking of electrical power consumption is very typical, particularly in sunny countries where the load is strongly influenced by air conditioning.

CPV has the highest efficiency and energy output, i.e., the lowest LCOE (levelized cost of electricity). Until now, most of the industry has been focusing on PV module cost. As a very new and immature industry, a great deal of the affects of scaling, productivity, and the cost reduction learning curve had to be achieved. Now, GW-level production has been achieved by top PV module suppliers, meaning that most of this learning has been integrated. So, cost reduction and efficiency improvement moves now into a continuous improvement mode, with a few percent maximum gain per year. Module cost reduction will be limited by raw materials cost contribution (glass, metals, semiconductor, etc.).

CPV has two fundamental cost advantages here: First of all, it minimizes the semiconductor content in the module thanks to high concentration. Any semiconductor material is far more expensive that any other material used in a PV solar plant. In addition, the higher the efficiency (CPV has 2X higher efficiency than multi-crystalline silicon technology), the better the usage of all the other material (glass, metal frame, etc.). That's why, for similar volumes, a CPV module will be cheaper than any other PV technology.

Polarizing Organic Photovoltaics

Polarizing organic photovoltaic (ZOPV) technology is demonstrated as a novel concept for energy harvesting and recycling technology. Novel, inverted quasi-bilayer device architecture is utilized to realize the ZOPV device. An anisotropic photovoltaic effect is obtained.

Operation

Up to three-fourths of the light energy wasted from LCD backlight illumination can be retrieved and utilized using polarizing organic photovoltaics. They can utilize external light energy also apart from backlight illumination using photovoltaic polarizers, which are present within the structure of the LCD screen.

Advantages

80% to 90% of the total energy utilized by any device with an LCD screen is used up by the backlight illumination. As polarizing organic photovoltaics can recycle up to 75% of wasted light energy, the efficiency of the device is increased.

Disadvantages

This simply incorporates additional conversion efficiency losses. These devices harvest their own light.

Cadmium Telluride Photovoltaics

Cadmium telluride photovoltaic or cadmium telluride thin film is a photovoltaic device that produces electricity from light by using a thin film of cadmium telluride (CdTe). CdTe solar cells differ from crystalline silicon photovoltaic technologies in that they use a smaller amount of semiconductor—a thin film—to convert absorbed light energy into electrons. Though CdTe solar cells are less efficient than crystalline silicon devices, they can be cheaper to produce, and the technology has the potential to surpass silicon in terms of cost per kilowatt of installed capacity. Although thin film technologies account for a small share of the market in photovoltaic devices, this segment is expected to grow rapidly, as there is much interest in developing novel manufacturing methods that could unlock economies of scale.

The first thin film technology to be developed was amorphous silicon, wherein silicon was randomly deposited onto a substrate (as opposed to the regular crystal lattice seen in wafer crystals). This technology had some problems: the process of depositing the silicon onto the substrate was time-consuming and costly, and the cells were inefficient. CdTe thin film technology is around 11 percent more efficient than amorphous silicon, as its band gap (the energy needed to excite an electron from its atom into a state where the electron can move freely) is 1.4 electron volts and thus matches the solar spectrum very well. It is also much more conducive to mass production, as the CdTe thin film can be deposited onto the substrate quickly and is a high-throughput technology. Each cellcomprises a junction of n-doped cadmium sulfide, known as the "window layer," on top of a p-doped layer of cadmium telluride, known as the "absorber." A transparent conductive front contact covers the cadmium sulfide, while the CdTe is in contact with a conductive rear surface substrate.

Despite its potential, the electronics industry has moved to try to remove elemental cadmium from personal electronics because cadmium is a cumulative poison. In Europe, the Restriction of Hazardous Substances (RoHS) legislation has been powerful in eliminating cadmium from electronic devices due to health effects. Not only does cadmium represent a health risk for consumers, but it is also dangerous for miners during extraction of the raw materials, for workers processing the material, and at end of life during disposal.

Proponents claim that cadmium in the form of a thin film solar cell is more stable and less soluble than in other electronics and that there would be little risk to health and the environment, as the alloys are encapsulated within the modules. However, there have been concerns regarding cadmium leaching from broken modules. Additionally, although it has been promoted that closed-loop recycling would address any concerns over end-of-life disposal, critics highlight that even closed-loop recycling systems do not recover everything.

Technology

Cell Efficiency

In August 2014, First Solar announced a device with 21.1% conversion efficiency. In February 2016, First Solar announced that they had reached a record 22.1% conversion efficiency in their CdTe cells. In 2014, the record module efficiency was also raised by First Solar from 16.1% up to 17.0%.

At this time, the company projected average production line module efficiency for its CdTe PV to be 17% by 2017, but by 2016, they predicted a module efficiency closer to ~19.5%.

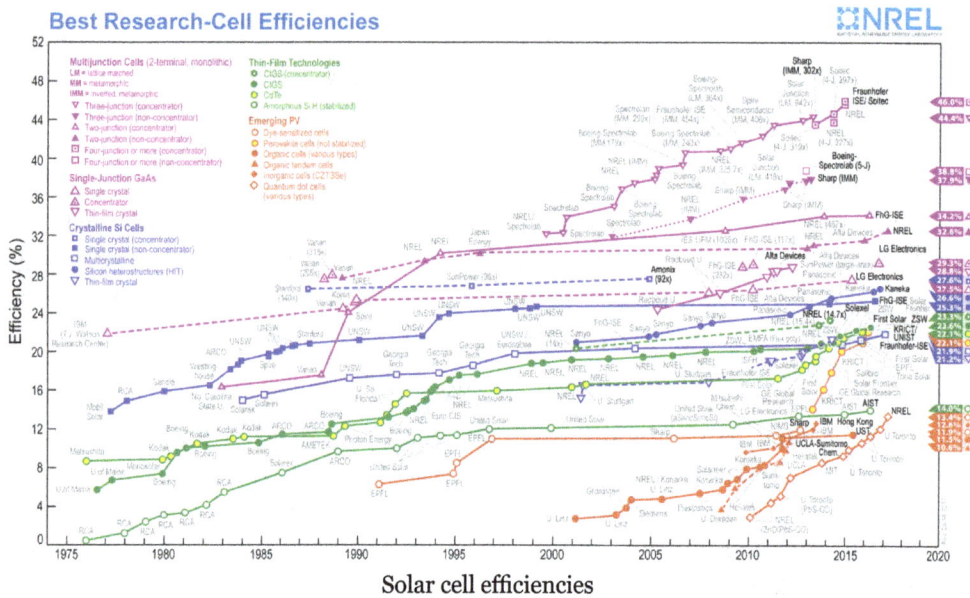

Solar cell efficiencies

Since CdTe has the optimal band gap for single-junction devices, efficiencies close to 20% (such as already shown in CIS alloys) may be achievable in practical CdTe cells.

Process Optimization

Process optimization improved throughput and lowered costs. Improvements included broader substrates (since capital costs scale sublinearly and installation costs can be reduced), thinner layers (to save material, electricity, and processing time), and better material utilization (to save material and cleaning costs). 2014 CdTe module costs were about $72 per 1 square metre (11 sq ft), or about $90 per module.

Ambient Temperature

Module efficiencies are measured in laboratories at standard testing temperatures of 25°C, however in the field modules are often exposed to much higher temperatures. CdTe's relatively low temperature coefficient protects performance at higher temperatures. CdTe PV modules experience half the reduction of crystalline silicon modules, resulting in an increased annual energy output of 5-9%.

Solar Tracking

Almost all thin film photovoltaic module systems to-date have been non-solar tracking, because module output was too low to offset tracker capital and operating costs. But relatively inexpensive single-axis tracking systems can add 25% output per installed watt. In addition, depending on the Tracker Energy Gain, the overall eco-efficiency of the PV system can be enhanced by lowering both system costs and environmental impacts. This is climate-dependent. Tracking also produces a smoother output plateau around midday, better matching afternoon peaks.

Materials

Cadmium

Cadmium (Cd), a toxic heavy metal considered a hazardous substance, is a waste byproduct of mining, smelting and refining sulfidic ores of zinc during zinc refining, and therefore its production does not depend on PV market demand. CdTe PV modules provide a beneficial and safe use for cadmium that would otherwise be stored for future use or disposed of in landfills as hazardous waste. Mining byproducts can be converted into a stable CdTe compound and safely encapsulated inside CdTe PV solar modules for years. A large growth in the CdTe PV sector has the potential to reduce global cadmium emissions by displacing coal and oil power generation.

Tellurium

Tellurium (Te) production and reserves estimates are subject to uncertainty and vary considerably. Tellurium is a rare, mildly toxic metalloid that is primarily used as a machining additive to steel. Te is almost exclusively obtained as a by-product of copper refining, with smaller amounts from lead and gold production. Only a small amount, estimated to be about 800 metric tons per year, is available. According to USGS, global production in 2007 was 135 metric tons. One gigawatt (GW) of CdTe PV modules would require about 93 metric tons (at current efficiencies and thicknesses). Through improved material efficiency and increased PV recycling, the CdTe PV industry has the potential to fully rely on tellurium from recycled end-of-life modules by 2038. In the last decade, new supplies have been located, e.g., in Xinju, China as well as in Mexico and Sweden. In 1984 astrophysicists identified tellurium as the universe's most abundant element having an atomic number over 40. Certain undersea ridges are rich in tellurium.

Cadmium Chloride/Magnesium Chloride

The manufacture of a CdTe cell includes a thin coating with cadmium chloride ($CdCl_2$) to increase the cell's overall efficiency. Cadmium chloride is toxic, relatively expensive and highly soluble in water, posing a potential environmental threat during manufacture. In 2014, research discovered that abundant and harmless magnesium chloride ($MgCl_2$) performs as well as cadmium chloride. This research may lead to cheaper and safer CdTe cells.

Safety

By themselves, cadmium and tellurium are toxic and carcinogenic, but CdTe forms a crystalline lattice that is highly stable, and is several orders of magnitude less toxic than cadmium. The glass plates surrounding CdTe material sandwiched between them (as in all commercial modules) seal during a fire and do not allow any cadmium release. All other uses and exposures related to cadmium are minor and similar in kind and magnitude to exposures from other materials in the broader PV value chain, e.g., to toxic gases, lead solder, or solvents (most of which are not used in CdTe manufacturing).

Recycling

Due to the exponential growth of photovoltaics the number of worldwide installed PV systems has increased significantly. First Solar established the first global and comprehensive recycling program in the PV industry in 2005. Its recycling facilities operate at each of First Solar's manufacturing plants and recover up to 95% of semiconductor material for reuse in new modules and 90% of glass for reuse in new glass products. A life cycle assessment of CdTe module recycling by the University of Stuttgart showed a reduction in primary energy demand in End-Of-Life from 81 MJ/m^2 to -12 MJ/m^2, a reduction of around 93 MJ/m^2, and in terms of global warming potential from 6 kg CO2-equiv./m^2 to -2.5 CO2-equiv./m^2, a reduction of around -8.5 CO2-equiv./m^2. These reductions show a highly beneficial change in the overall environmental profile of CdTe photovoltaic module. The LCA also showed that the main contributors to considered environmental impact categories are due to required chemicals and energy within the processing of CdTe modules.

Grain Boundaries

Grain boundary is the interface between two grains of a crystalline material and occurs when two grains meet. They are a type of crystalline defect. It is often assumed that the open-circuit voltage gap seen in CdTe, in comparison to both single crystal GaAs and the theoretical limit, may be in some way attributable to the grain boundaries within the material. There have however, been a number of studies which have suggested not only that GBs are not deleterious to performance but may in fact be beneficial as sources of enhanced carrier collection. So, the exact role of the grain boundaries in limitation of performance of CdTe-based solar cells remains unclear and the research is ongoing to address this question.

Photovoltaic Effect

The effect due to which light energy is converted to electric energy in certain semiconductor materials is known as photovoltaic effect. This directly converts light energy to electricity without any intermediate process. For demonstrating the photovoltaic effect let us assume a block of silicon crystal.

The upper portion of this block is doped with donor impurities and lower portion is doped with accept or impurities. Hence the concentration of free electrons is quite high in n-type region compared to p-type region and concentration of hole is quite high in p-type region compared to n-type region of the block. There will be a high concentration gradient of charge carriers across the junction line of the block. Free electrons from n-type region try to diffuse to p-type region and holes in p-type region try to diffuse to n-type region in the crystal. This is because charge carriers by nature always tend to diffuse from high concentration region to low concentration region. Each free electron of n-type region while comes to the p-type region due to diffusion, it leaves a positive donor ion behind it in the n-type region.

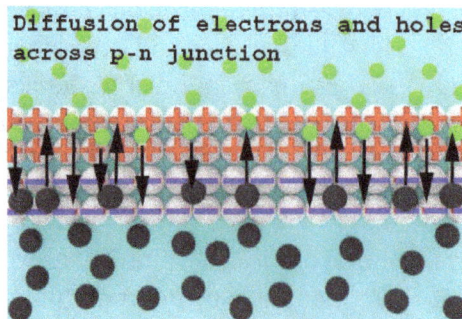

Diffusion of electrons and holes across p-n junction

This is because each of the free electrons in n-type region is contributed by one neutral donor atom. Similarly when a hole is diffused from p-type region to n-type region, it leaves a negative accept or ion behind it in p-type region.

Since each hole is contributed by one acceptor atom in p-type region. Both of these ions i.e. donor ions and acceptor ions are immobile and fixed at their position in crystal structure. It is needless to say that those free electrons of n-type region which are nearest to the p-type region first diffuse in the p-type region consequently create a layer of positive immobile donor ions in the n-type region adjacent to the junction.

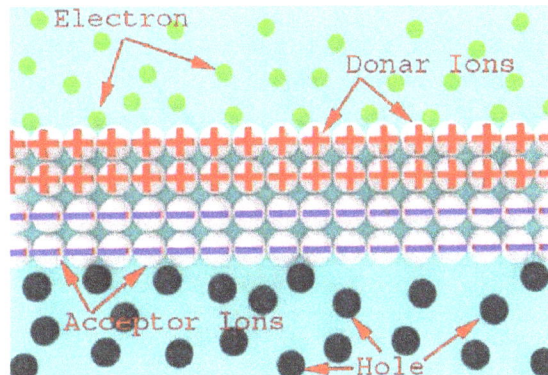

Similarly those free holes of p-type region which are nearest to the n-type region first diffuse in the n-type region consequently create a layer of negative immobile acceptor ions in the p-type region adjacent to the junction. These positive and negative ions concentration layer creates an electric field across the junction which is directed from positive to negative that in from n-type side to p-type side. Now due to presence of this electric field charge carriers in the crystal experience a force to drift according to the direction of this electric field. As we know the positive charge always drift in the direction of electric field hence the positively charged holes (if any) in n-type region now drift to the p-side of the junction.

On the other hand, negatively charged electrons in p-type region (if any) drift to n-region as negative charge always drift opposite to the direction of electric field. Across a p-n junction diffusion and drift of charge carriers continues. Diffusion of charge carriers creates and increases the thickness of the potential barrier across the junction and drift of the charge carriers reduces the thickness of the barrier. In normal thermal equilibrium condition and in absence of any external force, the diffusion of charge carrier is equal and opposite of drift of charge carriers hence the thickness of potential barrier remains fixed.

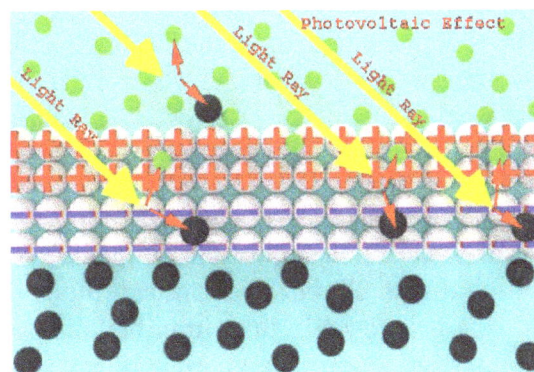

Now the n-type surface of the silicon crystal block is exposed to the sunlight. Some of the photons are absorbed by the silicon block. Some of the absorbed photon will have energy greater than the energy gap between valence and conduction band of valence electrons of the silicon atoms. Hence, some of the valence electrons in the covalent bond will be excited and jump out from the bond leaving behind a hole in the bond. In this way electron-hole pairs are generated in the crystal due to incident light. The holes of these light generated electron-hole pairs in the n-type side have enough probability of recombination with enormous electrons (majority carriers). Hence, solar cell is so designed, that the light- generated electrons or holes will not get enough chances to recombine with majority carriers.

The semiconductor (silicon) is so doped that the p-n junction forms in very close vicinity of exposed surface of the cell. If an electron hole pair is created within one minority carrier diffusion length, of the junction, the electrons of electron-hole pair will drift toward n-type region and hole of the pair will swept to p region due to in influence of electric field of the junction and hence on the average, it will contribute to current flow in an external circuit.

Two-photon Photovoltaic Effect

The two-photon photovoltaic (TPPV) effect is harvesting the energy of the photons lost to two-photon absorption (TPA). The effect is a nonlinear equivalent of the conventional (single-photon) photovoltaic effect widely used in solar cells. One possible application of the TPPV effect is self-powered electronic–photonic integrated circuits, in which the harvested energy can be used to drive electrical circuits on the same chip. A good example is self-powered in-line monitors (or optically powered remote sensors) in fiber-optic networks. Another potential application of the TPPV effect is remote power delivery to physical sensors installed in critical environments where electrical sparks are dangerous and copper cables must be avoided.

The TPPV effect was first demonstrated in silicon. Although the main purpose was to eliminate the nonlinear losses in silicon waveguides [TPA and free-carrier absorption (FCA)] by decreasing the carrier lifetime while, simultaneously, net electrical energy was harvested with reasonable power efficiency. Energy harvesting (or negative electrical power dissipation) based on the TPPV effect in silicon has been specifically investigated in Raman amplifiers, parametric wavelength converters, and optical modulators.

TPA has been long observed experimentally in gallium arsenide (GaAs) and the corresponding coefficient, β, reported in GaAs at 1.3 μm is 42.5 cm/GW, which is much higher than silicon's (3.3 cm/GW). At the telecommunication wavelength of 1.55 μm, β is reported to be around 15 cm/GW in GaAs compared with 0.7 cm/GW in silicon. Thus, the TPPV effect is expected to be stronger in GaAs. In this Letter, the TPPV effect is demonstrated in GaAs for the first time.

In principle, every two photons lost to TPA generate one electron–hole pair in the semiconductor material and these photogenerated carriers are available for photovoltaic conversion into electrical power. Figure below shows the TPA process at the two particular wavelengths studied here, i.e., 976 and 1550 nm. Figure below shows a simplified schematic on how nonlinear absorption along the waveguide (due to TPA at high optical intensities) is different than linear absorption at low intensities. FCA is ignored in this simplified diagram. Figure below illustrates how TPPV can be realized in a single-mode GaAs/AlGaAs waveguide using a *p-i-n* junction diode. The employed device resembles a standard edge-emitting laser but without any active (quantum well) region.

A theoretical model is developed to describe the TPPV effect in the GaAs/AlGaAs waveguide with vertical *p-i-n* hetero junction diodes, as shown in below figure Although a 01-D model is enough for simulation of conventional solar cells, a 2-D approach that considers the optical intensity distribution of the guided mode is applied here. The numerical model is developed in COMSOL's Multi physics module. The model solves the drift-diffusion equations of carrier transport in semiconductors in terms of the quasi-Fermi potentials along with Poisson's equation.

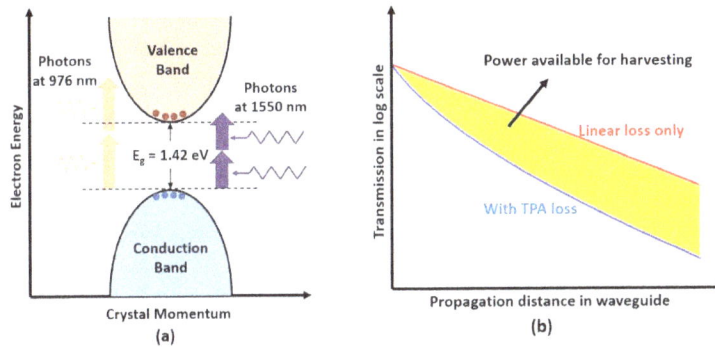

Figure: (a) TPA in GaAs at studied wavelengths of 976 and 1550 nm and (b) waveguide loss with and without TPA. The carriers generated in GaAs by TPA are in principle available for photovoltaic conversion

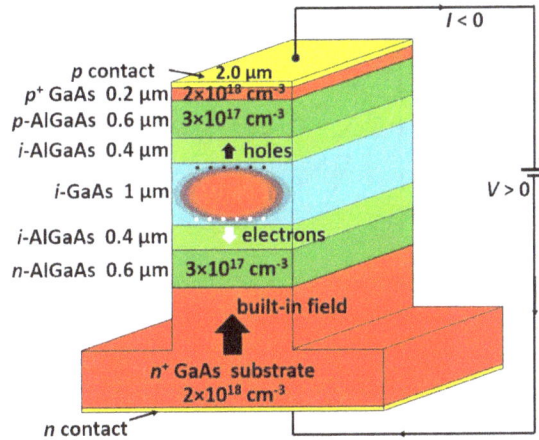

Figure: Schematic of a GaAs/AlGaAs waveguide diode for photovoltaic power conversion using the process

The Shockley–Read–Hall recombination is taken into account in this model assuming the trap energy level is located at the middle of the bandgap. The electron and hole bulk recombination lifetimes, τ_n and τ_p, in bulk GaAs are of the order of 10^{-8} s, about 2 orders of magnitude smaller than those in bulk silicon. In GaAs/AlGaAs waveguides, the effective free-carrier lifetime τ_{eff} can be even smaller when the surface recombination at the waveguide sidewalls dominates over the bulk recombination. τ_{eff} of 250 ps is reported for a 2.4 μm × 0.8 μm GaAs/Al$_{0.8}$Ga$_{0.2}$ as ridge waveguide . In this work, τ_{eff} is assumed to be 100 ps, as good agreement between numerical simulation and experimental data is achieved with this value figures. This low value suggests that the lifetime is dominated by surface rather than bulk recombination. This value is reasonable considering the sidewall roughness induced by the dry etching process, which may have increased the surface recombination. Surface recombination reduces the power efficiency of the TPPV effect as the electrons and holes recombine before they are collected at the contacts. Higher τ_{eff} can be achieved by optimizing the dry etching recipe and smoothening the sidewalls of the waveguide. Table summaries the other parameters and material properties used in the simulation.

From the current density, J, calculated by COMSOL the total current is calculated by integrating J over the waveguide length, L. Finally, the ohmic loss of the electrodes and the contacts is included as a series resistance, R_s, in the circuit. The ohmic loss will reduce the harvested electrical power and might possibly lower the short-circuit current if R_s is excessively high.

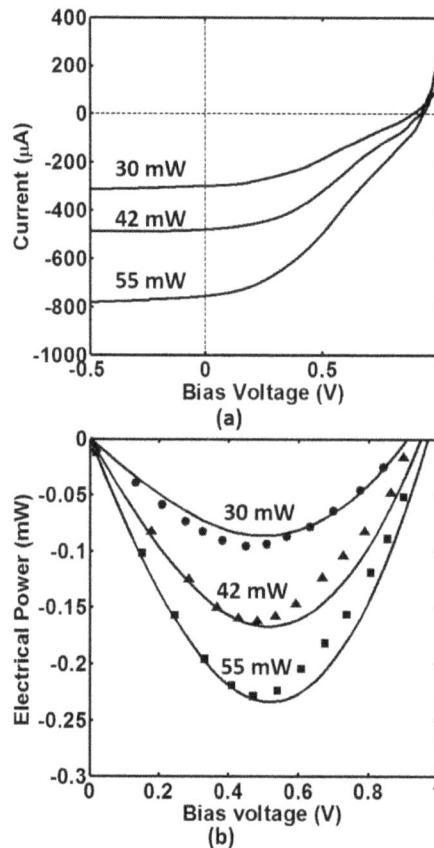

Figure: (a) I–V characteristics of the diodes at wavelength of 976 nm for three different coupled input powers and (b) the corresponding P–V characteristics of the diodes from numerical simulation (solid line) and experiment (circles, triangles, and squares).

It should be noted that the waveguide in above figure is multimode at both wavelengths studied in this work. As the carrier photogeneration rate is proportional to the square of the optical intensity, the intensity distribution across the cross section of the waveguide is not expected to be important as far as the mode and the coupled input optical power are well confined in the GaAs core layer. It was confirmed by simulation that, for the same coupled input power, the change in the collected current density is negligible when the intensity distribution of fundamental and higher orders modes are compared. This can also be explained mathematically by a simplified 1-D model for slab waveguides, assuming the electric field in the waveguide core follows a cosine function and drops to= zero at the core–cladding interface.

Table: Material Properties Used in This Study

Parameters		Unit	GaAs	$Al_{0.15}Ga_{0.85}As$
Relative permittivity ε_r 976 nm		-	12.42	11.88
	1550 nm	-	11.41	11.27
Intrinsic carrier density n_i		cm^{-3}	2.2×10^6	6.6×10^4
Electron mobility μ_n		$cm^2/(Vs)$	8500	4925
Hole mobility μ_p		$cm^2/(Vs)$	400	241
Effective mass of electron m_e		m_0	0.067	0.076
Effective mass of hole m_h		m_0	0.48	0.548
Energy bandgap E_g		eV	1.42	1.611
Electron affinity X		eV	4.06	3.905

The heterostructure in above figure was grown by molecular beam epitaxy and several devices with identical widths of 2.0 µm but varying etch depths were fabricated using standard optical lithography, lift-off, and dry etching processes. The length of all the waveguides reported here is 4.5 mm.

For characterization of the devices, optical energy is coupled into the intrinsic GaAs layer (waveguide core) through a lensed fiber. The coupling loss is estimated to be around 6 dB. Two wavelengths, one closer to GaAs's bandgap (from a high-power 976 nm diode laser) and 1550 nm (from an erbium-doped fiber amplifier) were studied. The choice of these wavelengths is merely based on the high-power sources available to us.

The linear propagation loss, α, of the waveguide is measured by the cut-back method and is estimated to be 7 dB/cm at 976 nm and around 18 dB/cm at 1550 nm. The high linear loss at 1550 nm can be explained by the mode leakage into the n^+-GaAs substrate as the AlGaAs bottom cladding layer is only 1 µm thick, as confirmed by the imaginary part of simulated effective index. To reduce the mode leakage and increase the power efficiency of the device at this longer wavelength (1550 nm), the bottom cladding layer has to be thicker than 2 µm. The current–voltage (I–V) characteristics of the fabricated waveguide diodes were measured with a curve tracer at various coupled pump powers and wavelengths.

Figure above shows the measured I–V characteristics at 976 nm at three different coupled input powers. The etch depth of the ridge waveguide is 4.2 µm (all the way down to the n$^+$-GaAs substrate). The TPPV effect is clearly observed when the device is biased in the fourth quadrant of its I–V characteristics, i.e., the carriers generated by TPA are swept out by the built-in field of the p-i-n junction and collected at the electrodes. The measured current is a combination of the photocurrent (from TPA), the minority carrier diffusion current, and the recombination current (including both bulk and surface recombination).

The power–voltage (P–V) characteristics of the diode at three different coupled input powers are shown in above figure. Evidently, 230 µW of electrical power can be scavenged from this device. The solid lines show agreeable simulation results based on the model described before. R_s and β are treated as fitting parameters in the simulation and are estimated to be around 700 Ω and 40 cm/GW, respectively. The β value obtained here is comparable to those reported in.

It should be mentioned that the device has also been tested before annealing. At pump power of 55 mW, the generated electrical power is only half of that harvested from an annealed device. This indicates that annealing alleviates the ohmic loss remarkably and is crucial in fabrication of the photovoltaic devices. Up to 9% power efficiency, excluding the coupling loss, is theoretically predicted in long (several centimeters), low-loss GaAs waveguides. The rather low wall-plug efficiency observed is attributed to the high series resistance at the contacts, as well as fabrication imperfections, which result in high linear propagation loss and strong surface recombination at the sidewalls of the waveguide.

Figure below shows the $I-V$ characteristics of a similar device at 976 nm, but with a shorter etch depth of 1.0 μm (down to the center of the AlGaAs cladding layer). The shallow-etched device has two advantages. First, since the GaAs core layer is not fully etched, the scattering loss of the optical mode is less and the surface recombination is weakened; second, the selectivity of the applied dry etching process between GaAs and gold is around 20:1. Therefore, the top contact gold layer is around 150 μm thicker than that of the deep-etched device due to less etching time in the process, which decreases R_s by about 200 Ω. However, the generated photocurrent and the harvested electrical power are less than those in the deepetched device because of the existence of a huge slab mode, which results in low optical intensity in the waveguide core, i.e., the photons generated in the slab cannot be efficiently collected by the diode. Figure below presents the measured electrical power harvested and delivered to a load resistance of 1 kΩ, for the two studied etch depths.

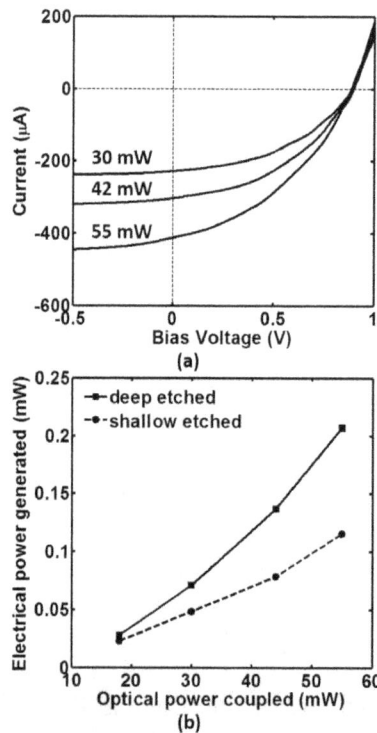

Figure: (a) Measured I–V characteristics of the shallow-etched devices at wavelength of 976 nm for three different coupled input powers. (b) Measured electrical power generated in a 1 kΩ load resistance for both etch depths

Next, TPPV in GaAs was studied at the telecommunication wavelength of 1550 nm. The etch depth of the ridge waveguide is 4.2 μm in this case. The $I-V$ and $P-V$ characteristics measured at three different coupled input powers are shown in figures. Once again, good agreement between the numerical and experimental results is observed. β is estimated to be 17 cm/GW at this wavelength, in accordance with the value reported and is still much higher than that in silicon (0.7 cm/GW at 1550 nm). The maximum scavenged electrical power is −220 μW.

Finally, to predict the ultimate performance limit of GaAs TPPV power converters, devices with more idealistic, but reasonable, conditions were simulated. Negligible ohmic loss of the contacts, as well as low linear propagation loss of 1 dB/cm are assumed in this simulation. Such low α is achievable in micrometer-sized waveguides if the bottom cladding layer is designed to be thicker and the fabrication processes, especially the dry etching recipe, are optimized.

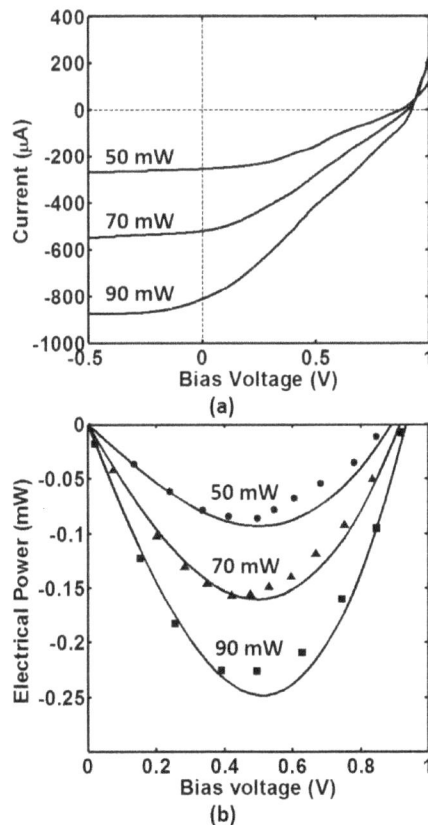

Figure: (a) I–V characteristics of the diodes at wavelength of 1550 nm for three different coupled input powers and (b) the corresponding P–V characteristics of the diodes from numerical simulation (solid line) and experiment (circles, triangles, and squares).

Figure below presents the simulated maximum possible generated electrical power versus input power for three different waveguide lengths at 1550 nm. 12 mW of electrical power can be harvested at input power of 150 mW in a 5-cm-long device, i.e., a power efficiency of 8%. At higher input power, a slight increase in the slope is clearly visible in all three line. This does not occur in silicon due to the high FCA loss in silicon waveguides. Further increasing the length of the present GaAs waveguide can only slightly improve the power efficiency as the optical intensity is not high enough after the light propagates a certain distance along the waveguide and gets absorbed gradually.

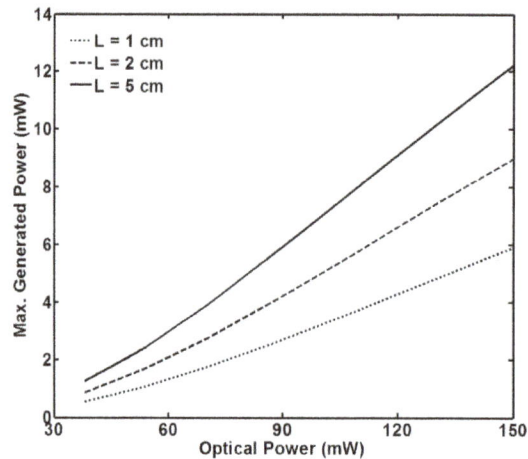

In conclusion, the TPPV effect is experimentally demonstrated in GaAs for the first time at wavelengths of 976 and 1550 nm. Good agreement between the simulation and experimental results is observed. The TPPV effect is more efficient at 976 nm due to larger β. A maximum electrical power of 230 μW is generated with 90 mW coupled input power at 1550 nm. Higher power efficiency at 1550 nm can be achieved by optimizing the waveguide design, particularly by increasing the thickness of the intrinsic bottom cladding layer. Power efficiency up to 8% is theoretically predicted in a 5-cm-long device at input power of 150 mW, which is higher than those achievable in silicon.

Photoelectric Effect

The photoelectric effect is a quantum electronic phenomenon in which electrons are emitted from matter after the absorption of energy from electromagnetic radiation such as x-rays or visible light. The emitted electrons can be referred to as *photoelectrons* in this context. The effect is also termed the *Hertz Effect,* based on its discovery by Heinrich Hertz, although the term has generally fallen out of use.

Study of the photoelectric effect led to important steps in understanding the quantum nature of light and electrons and influenced the formation of the concept of wave–particle duality.

The term may also refer to the photoconductive effect (also known as photoconductivity or photoresistivity), the photovoltaic effect, or the photoelectrochemical effect.

When a metallic surface is exposed to electromagnetic radiation above a certain threshold frequency, the light is absorbed and electrons are emitted. In 1902, Philipp Eduard Anton von Lenard observed that the energy of the emitted electrons increased with the frequency, or color, of the light. This was at odds with James Clerk Maxwell's wave theory of light, which predicted that the energy would be proportional to the intensity of the radiation. In 1905, Albert Einstein solved this paradox by describing light as composed of discrete quanta, now called photons, rather than continuous waves. Based upon Max Planck's theory of black-body radiation, Einstein theorized that the energy in each quantum of light was equal to the frequency multiplied by a constant, later called Planck's constant. A photon above a threshold frequency has the required energy to eject a single electron, creating the observed effect. This discovery led to the quantum revolution in physics and earned Einstein the Nobel Prize in 1921.

Explanation

The photons of the light beam have a characteristic energy determined by the frequency of the light. In the photoemission process, if an electron absorbs the energy of one photon and has more energy than the work function (the electron binding energy), it is ejected from the material. If the photon energy is too low, the electron is unable to escape the surface of the material. Increasing the intensity of the light beam increases the number of photons in the light beam, and thus increases the number of electrons emitted without increasing the energy that each electron possesses. Thus the energy of the emitted electrons does not depend on the intensity of the incoming light, but only on the energy of the individual photons.

Electrons can absorb energy from photons when irradiated, but they follow an "all or nothing" principle. All of the energy from one photon must be absorbed and used to liberate one electron from atomic binding, or the energy is re-emitted. If the photon energy is absorbed, some of the energy liberates the electron from the atom, and the rest contributes to the electron's kinetic energy as a free particle.

Experimental Results of the Photoelectric Emission

1. For a given metal and frequency of incident radiation, the rate at which photoelectrons are ejected is directly proportional to the intensity of the incident light.

2. For a given metal, there exists a certain minimum frequency of incident radiation below which no photoelectrons can be emitted. This frequency is called the threshold frequency.

3. Above the threshold frequency, the maximum kinetic energy of the emitted photoelectron is independent of the intensity of the incident light but depends on the frequency of the incident light.

4. The time lag between the incidence of radiation and the emission of a photoelectron is very small, less than 10^{-9} second.

Equations

In effect quantitatively using Einstein's method, the following equivalent equations are used:

Energy of photon = Energy needed to remove an electron + Kinetic energy of the emitted electron

Algebraically:

$$hf = \phi + E_{k_{max}}$$

where

-   h is Planck's constant,
-   f is the frequency of the incident photon,

- $\phi = hf_0$ is the work function (sometimes denoted W instead), the minimum energy required to remove a delocalized electron from the surface of any given metal,

- $E_{k_{max}} = \frac{1}{2}mv_m^2$ is the maximum kinetic energy of ejected electrons,

- f_0 is the threshold frequency for the photoelectric effect to occur,

- m is the rest mass of the ejected electron, and

- v_m is the speed of the ejected electron.

Since an emitted electron cannot have negative kinetic energy, the equation implies that if the photon's energy (hf) is less than the work function (ϕ), no electron will be emitted.

According to Einstein's special theory of relativity the relation between energy (E) and momentum (p) of a particle is:

$E = \sqrt{(pc)^2 + (mc^2)^2}$, where m is the rest mass of the particle and c is the velocity of light in a vacuum.

Three-step Model

The photoelectric effect in crystalline material is often decomposed into three steps:

1. Inner photoelectric effect: The hole left behind can give rise to auger effect, which is visible even when the electron does not leave the material. In molecular solids, photons are excited in this step and may be visible as lines in the final electron energy. The inner photoeffect has to be dipole allowed. The transition rules for atoms translate via the tight-binding model onto the crystal. They are similar in geometry to plasma oscillations in that they have to be transversal.

2. Ballistic transport of half of the electrons to the surface: Some electrons are scattered.

3. Electrons escape from the material at the surface.

In the three-step model, an electron can take multiple paths through these three steps. All paths can interfere in the sense of the path integral formulation. For surface states and molecules the three-step model does still make some sense as even most atoms have multiple electrons which can scatter the one electron leaving.

Uses and Effects

Photodiodes and Phototransistors

Solar cells (used in solar power) and light-sensitive diodes use a variant of the photoelectric effect, but not ejecting electrons out of the material. In semiconductors, light of even relatively low energy, such as visible photons, can kick electrons out of the valence band and into the higher-energy conduction band, where they can be harnessed, creating electric current at a voltage related to the bandgap energy.

Image Sensors

Video camera tubes in the early days of television used the photoelectric effect; newer variants used photoconductive rather than photoemissive materials.

Silicon image sensors, such as charge-coupled devices, widely used for photographic imaging, are based on a variant of the photoelectric effect, in which photons knock electrons out of the valence band of energy states in a semiconductor, but not out of the solid itself.

Gold-leaf Electroscope

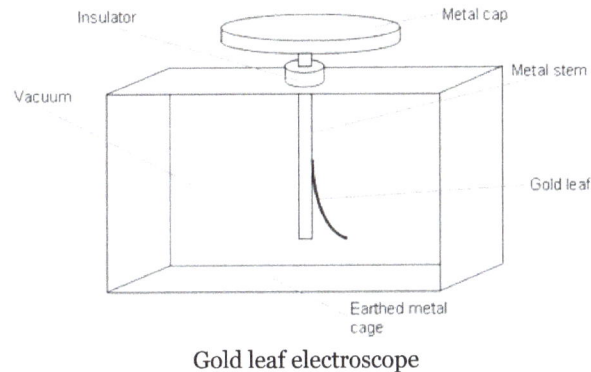

Gold leaf electroscope

Gold-leaf electroscopes are designed to detect static electricity. Charge placed on the metal cap spreads to the stem and the gold leaf of the electroscope. Because they then have the same charge, the stem and leaf repel each other. This will cause the leaf to bend away from the stem.

The electroscope is an important tool in illustrating the photoelectric effect. Let us say that the scope is negatively charged throughout. There is an excess of electrons and the leaf is separated from the stem. But if we then shine high-frequency light onto the cap, the scope discharges and the leaf will fall limp. This is because the frequency of the light shining on the cap is above the cap's threshold frequency. The photons in the light have enough energy to liberate electrons from the cap, reducing its negative charge. This will discharge a negatively charged electroscope and further charge a positive electroscope.

However, if the EM radiation hitting the metal cap does not have a high enough frequency, (its frequency is below the threshold value for the cap) then the leaf will never discharge, no matter how long one shines the low-frequency light at the cap.

Photoelectron Spectroscopy

Since the energy of the photoelectrons emitted is exactly the energy of the incident photon minus the material's work function or binding energy, the work function of a sample can be determined by bombarding it with a monochromatic X-ray source or UV source (typically a helium discharge lamp), and measuring the kinetic energy distribution of the electrons emitted.

Photoelectron spectroscopy is done in a high vacuum environment, since the electrons would be scattered by air.

A typical electron energy analyzer is a concentric hemispherical analyzer (CHA), which uses an electric field to divert electrons different amounts depending on their kinetic energies. For every element and core (atomic orbital) there will be a different binding energy. The many electrons created from each will then show up as spikes in the analyzer, and can be used to determine the elemental composition of the sample.

Spacecraft

The photoelectric effect will cause spacecraft exposed to sunlight to develop a positive charge. This can get up to the tens of volts. This can be a major problem, as other parts of the spacecraft in shadow develop a negative charge (up to several kilovolts) from nearby plasma, and the imbalance can discharge through delicate electrical components. The static charge created by the photoelectric effect is self-limiting, though, because a more highly-charged object gives up its electrons less easily.

Moon Dust

Light from the sun hitting lunar dust causes it to become charged through the photoelectric effect. The charged dust then repels itself and lifts off the surface of the Moon by electrostatic levitation. This manifests itself almost like an "atmosphere of dust," visible as a thin haze and blurring of distant features, and visible as a dim glow after the sun has set. This was first photographed by the Surveyor program probes in the 1960s. It is thought that the smallest particles are repelled up to kilometers high, and that the particles move in "fountains" as they charge and discharge.

Night Vision Devices

Photons hitting a gallium arsenide plate in night vision devices cause the ejection of photoelectrons due to the photoelectric effect. These are then amplified into a cascade of electrons that light up a phosphor screen.

References

- Chaves, Julio (2015). Introduction to Nonimaging Optics, Second Edition. CRC Press. ISBN 978-1482206739.

- Sun-free-photovoltaics-0728: news.mit.edu, Retrieved 25 June 2018

- Andrews, Rob W.; Pollard, Andrew; Pearce, Joshua M., "Photovoltaic system performance enhancement with non-tracking planar concentrators: Experimental results and BDRF based modelling," Photovoltaic Specialists Conference (PVSC), 2013 IEEE 39th, pp.0229,0234, 16–21 June 2013. doi:10.1109/PVSC.2013.6744136

- Concentrator-photovoltaics: renewableenergyworld.com, Retrieved 31 March 2018

- "BP solar ditches thin-film photovoltaics - IEEE Journals & Magazine". ieeexplore.ieee.org. Retrieved 2018-05-20.

- Cadmium-telluride-solar-cell, technology: britannica.com, Retrieved 28 May 2018

- Kinsey, G. S.; Bagienski, W.; Nayak, A.; Liu, M.; Gordon, R.; Garboushian, V. (2013-04-01). "Advancing Efficiency and Scale in CPV Arrays". IEEE Journal of Photovoltaics. 3 (2): 873–878. doi:10.1109/JPHOTOV.2012.2227992. ISSN 2156-3381.

- What-is-photovoltaic-effect: electrical4u.com. Retrieved 11 July 2018

- Friedman, Thomas L. (5 November 2009). Hot, Flat, and Crowded: Why The World Needs A Green Revolution - and How We Can Renew Our Global Future. Penguin Books Limited. p. 388. ISBN 978-0-14-191850-1.

- Photoelectric-effect: newworldencyclopedia.org, Retrieved 20 March 2018

- Cole, IR, Betts, TR, Gottschalg, R (2012), "Solar profiles and spectral modeling for CPV simulations", IEEE Journal of Photovoltaics, 2 (1): 62–67, doi:10.1109/JPHOTOV.2011.2177445, ISSN 2156-3381

2

Photovoltaic Materials

Among the semiconducting materials used in photovoltaics, silicon crystals are the most common. The topics elaborated in this chapter will help in developing a better perspective about the materials that are used in photovoltaics, such as crystalline silicon, polycrystalline silicon, monocrystalline silicon, cadmium telluride, etc.

Semiconductor Materials

The PV cell is composed of semiconductor material, which combines some properties of metals and some properties of insulators. That makes it uniquely capable of converting light into electricity. When light is absorbed by a semiconductor, photons of light can transfer their energy to electrons, allowing the electrons to flow through the material as electrical current. This current flows out of the semiconductor to metal contacts and then makes its way out to power your home and the rest of the electric grid.

A semiconductor materials within a photovoltaic system can vary from silicon, polycrystalline thin films, or single-crystalline thin films. The silicon materials include single-crystalline silicon, multi-crystalline silicon, and amorphous silicon. Single-crystalline silicon has a regular structure, allowing a better response rate compared to that seen in the multi-crystalline structure.

Amorphous silicon is comprised of atoms that are randomly placed together, exhibiting a lower response rate than that observed in the single-crystalline structure. Capable of capturing more light than the crystalline silicone, alloying amorphous silicon with germanium or carbon can intensify this property.

Copper indium diselenide (CIS), cadmium telluride (CdTe), and thin-film silicon are certain polycrystalline thin film materials often used, whereas high-efficiency material such as gallium arsenide (GaAs) often comprise single-crystalline thin film materials. Each of these materials

possesses unique strengths and characteristics that influence its suitability for the desired application. Of these traits include the material's crystallinity, band-gap, absorption, and manufacturing complexity.

Crystallinity

The crystallinity of a material indicates how perfectly ordered the atoms are in the crystal structure. Silicon, as well as other solar cell semiconductor materials, comes in various forms, including:

- Single-crystalline,
- Multicrystalline,
- Polycrystalline, and
- Amorphous.

In a single-crystal material, the atoms that make up the framework of the crystal are repeated in a very regular, orderly manner from layer to layer. In contrast, in a material composed of numerous smaller crystals, the orderly arrangement is disrupted moving from one crystal to another.

Bandgap

The bandgap of a semiconductor material is the: minimum energy needed to move an electron from its bound state within an atom to a free state.

This free state is where the electron can be involved in conduction.

The lower energy level of a semiconductor is called the valence band, and the higher energy level where an electron is free to roam is called the conduction band.

The bandgap (often symbolized by Eg) is the energy difference between the conduction and valence bands.

Absorption

The absorption coefficient of a material indicates how far light with a specific wavelength (or energy) can penetrate the material before being absorbed. A small absorption coefficient means that light is not readily absorbed by the material.

The absorption coefficient of a solar cell depends on two factors:

1. Material of the cell,
2. Wavelength or energy of the light being absorbed.

Solar cell material has an abrupt edge in its absorption coefficient because light with energy below the material's bandgap cannot free an electron.

External Influences on the Semiconductor

Determined based on order arrangement of atoms within the crystal structure, the crystallinity of a semiconductor can affect the charge transport, current density, and power conversion efficiency within a solar cell. The bandgap of a semiconductor material refers to the minimum energy required to move an electron from its bound state to a free state, allowing the conduction of the electron to occur. The valence band of a semiconductor is the lower energy level, and the conduction band is the higher energy level.

Often symbolized by Eg, the band-gap describes the energy difference between the valence and conduction bands. Indicated by how far light with a specific wavelength can penetrate the material, the absorption coefficient determines the ability of light to be absorbed by the material. This coefficient depends on the material of the cell and the wavelength of the light being absorbed.

The cost and manufacturing complexity of the wide variety of materials and devices for semiconductor materials is based on many factors, including the amount and type of material used, the time required for production, and the movement of cells into different deposition chambers within the module. Each of these characteristics plays a significant role in determining the capability of a photovoltaic system to efficiently produce energy for a given application.

Commonly used Material in Different Kind of Photovoltaics

Silicon

Silicon is, by far, the most common material used in solar cells, representing approximately 90% of the modules sold today. It is also the second most abundant material on Earth (after oxygen) and the most common semiconductor used in computer chips. Crystalline silicon cells are made of silicon atoms connected to one another to form a crystal lattice. This lattice provides an organized structure that makes conversion of light into electricity more efficient.

Solar cells made out of silicon currently provide a combination of high efficiency, low cost, and long lifetime. Modules are expected to last for 25 years or more, still producing more than 80% of their original power after this time.

Thin-film Photovoltaics

A thin-film solar cell is made by depositing one or more thin layers of PV material on a supporting material such as glass, plastic, or metal. There are two main types of thin-film PV semiconductors on the market today: cadmium telluride (CdTe) and copper indium gallium diselenide (CIGS). Both materials can be deposited directly onto either the front or back of the module surface.

CdTe is the second-most common PV material after silicon and enables low-cost manufacturing processes. While this makes them a cost-effective alternative, their efficiencies still aren't quite as high. CIGS cells have favorable electronic and optical properties, though the complexity involved in combining four elements makes the transition from lab to manufacturing or challenging. Both CdTe and CIGS require more protection than silicon to enable long-lasting operation outdoors.

Organic Photovoltaics

Organic PV, or OPV, cells are composed of carbon-rich polymers and can be tailored to enhance a specific function of the cell, such as sensitivity to a certain type of light. This technology has the theoretical potential to provide electricity at a lower cost than silicon or thin-film technologies. OPV cells are only about half as efficient as crystalline silicon and have shorter operating lifetimes, but could be less expensive to manufacture in high volumes. They can also be applied to a variety of supporting materials, making OPV able to serve a wide variety of uses.

Concentration Photovoltaics

Concentration PV, also known as CPV, focuses sunlight onto a solar cell by using a mirror or lens. By focusing sunlight onto a small area, less PV material is required. PV materials become more efficient at energy conversion as the light becomes more concentrated, so the highest overall efficiencies are obtained with CPV cells and modules. However, more expensive materials, manufacturing techniques, and tracking are required, so demonstrating the necessary cost advantage over today's high-volume silicon modules has become challenging.

Crystalline Silicon

Crystalline silicon is the most widely used semiconductor material for photovoltaic (PV) conversion. Close to 80 % of all PV modules are crystalline silicon based. Although the technology is mature with respect to production yield and with respect to reliability, there is still a large potential in increasing the efficiency and in reducing the cost.

There are two types of crystalline silicon solar cells used in crystalline silicon photovoltaics:

- Mono-crystalline silicon, produced by slicing wafers from a high-purity single crystal ingot,

- Multi-crystalline silicon, made by sawing a cast block of silicon first into bars and then into wafers.

Mono-crystalline silicon solar cells have higher efficiencies than multi-crystalline silicon solar cells.

Production

Typical crystalline silicon solar cells are produced from monocrystalline (single-crystal) silicon or multicrystalline silicon. Monocrystalline cells are produced from pseudo-square silicon wafers, substrates cut from boules grown by the Czochralski process, the float-zone technique, ribbon growth, or other emerging techniques. Multicrystalline silicon solar cells are traditionally made from square silicon substrates cut from ingots cast in quartz crucibles.

To reduce the amount of light reflected by the solar cell—and therefore not used to generate current—an antireflective coating (ARC), often titanium dioxide (TiO_2) or silicon nitride (SiN), is deposited on the silicon surface. To increase light trapping and absorption, the top of the solar cell can be textured with micrometer-sized pyramidal structures, formed by a chemical etch process.

To create a p-n junction, typically a phosphorus-doped n^+ region is created on top of a boron-doped p-type silicon substrate. A metal electrode, such as aluminum, forms the back contact, whereas the front contact is most often formed using screen-printed silver paste applied on the top of the ARC layer.

Charge-carrier collection in a crystalline silicon solar cell is achieved by minority-carrier diffusion within the p-doped and n-doped layers. Long diffusion lengths (> 200 micrometers) assist carrier collection over the entire range of the solar cell thickness where the optical absorption occurs.

Energy Payback Time

The energy payback time (EPBT) describes the time span a PV system needs to operate in order to generate the same amount of energy that was used for its manufacture and installation. This energy amortization, given in years, is also referred to as *break-even* energy payback time. The EPBT depends vastly on the location where the PV system is installed (e.g. the amount of sunlight available) and on the efficiency of the system, namely the type of PV technology and the system's components.

In life-cycle analysis (LCA) from the 1990s, the energy payback time had often been cited to be as high as 10 years. Although the time span already decreased to less than 3 years in the early 2000s, the myth that "solar PV does not pay back the energy used to create it" seems to persists up to the present day.

The EPBT relates closely to the concepts of net energy gain (NEG) and energy returned on energy invested (EROI). They are both used in energy economics and refer to the difference between the energy expended to harvest an energy source and the amount of energy gained from that harvest. The NEG and EROI also take the operating lifetime of a PV system into account and an effective production life of 25 to 30 years is typically assumed, as many manufacturers now provide a 25-year warranty on their products. From these metrics, the Energy payback Time can be derived by calculation.

Energy Payback Time in Years for different locations and technologies

Location	Crystalline Silicon		Thin-film			CPV	Radiation Map Color	Global Solar Potential in kWh/m²/a
Examples	Mono	Multi	a-Si	CIGS	CdTe			
North-and Central Europe, Canada	3.3	2.1	2.4	1.7	1.1	–	1200 kWh	
Southern Europe, USA, South America, India	1.8	1.2	1.3	0.9	0.7	0.8	1700 kWh	
American Southwest, Australia, Africa, Middle East	1.5	<1.2	0.9	<0.9	<0.7	<0.8	1900 kWh	

EPBT Improvements

The EPBT has always been longer for PV systems using crystalline silicon than thin-film technology. This is due to the fact, that silicon is produced by the reduction of high-grade quartz sand in electric furnaces. This carbo-thermic smelting process occurs at high temperatures of more than 1000°C and is very energy intensive, using about 11 kilowatt-hours (kWh) per produced kilogram of silicon. However, the energy payback time has shortened significantly over the last years, as crystalline silicon cells became ever more efficient in converting sunlight, while the thickness of the wafer material was constantly reduced and therefore required less silicon for its manufacture. Within the last ten years, the amount of silicon used for solar cells declined from 16 to 6 grams per watt-peak. In the same period, the thickness of a c-Si wafer was reduced from 300 μm, or microns, to about 160–190 μm. Crystalline silicon wafers are nowadays only 40 percent as thick as they used to be in 1990, when they were around 400 μm. The sawing techniques that slice crystalline silicon ingots into wafers have also improved by reducing the kerf loss and making it easier to recycle the silicon sawdust.

Key Parameters for Material and Energy Efficiency

Parameter	Mono-Si	CdTe
Cell efficiency	16.5%	15.6%
Derate cell to module efficiency	8.5%	13.9%
Module efficiency	15.1%	13.4%
Wafer thickness / layer thickness	190 μm	4.0 μm
Kerf loss	190 μm	–
Silver per cell	9.6 g/m²	–
Glass thickness	4.0 mm	3.5 mm
Operational lifetime	30 years	30 years

Toxicity

With the exception of amorphous silicon, most commercially established PV technologies use toxic heavy metals. CIGS often uses a CdS buffer layer, and the semiconductor material of CdTe-technology itself contains the toxic cadmium (Cd). In the case of crystalline silicon modules, the solder material, that joins together the copper strings of the cells, contains about 36 percent of lead (Pb). Moreover, the paste used for screen printing front and back contacts contains traces of Pb and sometimes Cd as well. It is estimated, that about 1,000 metric tonnes of Pb have been used for 100 gigawatts of c-Si solar modules. However, there is no fundamental need for lead in the solder alloy.

Silver

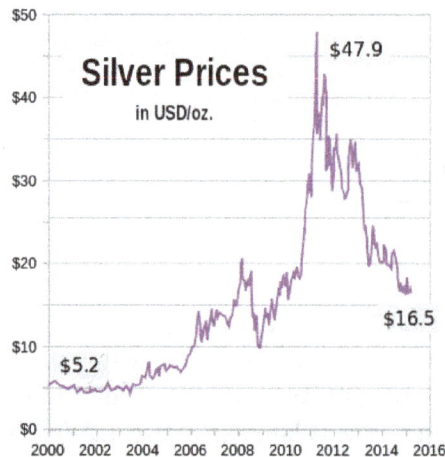

Silver price chart since 2000

Cell Technologies

PERC Solar Cell

Passivated emitter rear contact (PERC) solar cells consists in the addition of an extra layer to the rear-side of a solar cell. This dielectric passive layer acts to reflect unabsorbed light back to the solar cell for a second absorption attempt increasing the solar cell efficiency.

A PERC is created through an additional film deposition and etching process. Etching can be done either by chemical or laser processing.

HIT Solar Cell

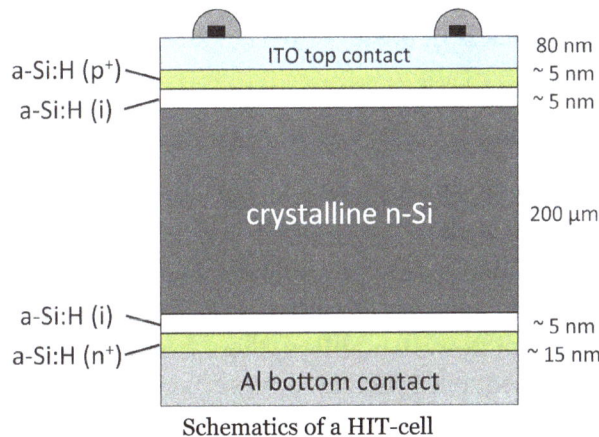

Schematics of a HIT-cell

A HIT solar cell is composed of a mono thin crystalline silicon wafer surrounded by ultra-thin amorphous silicon layers. The acronym HIT stands for Heterojunction with Intrinsic Thin layer. HIT cells are produced by the Japanese multinational electronics corporation Panasonic . Panasonic and several other groups have reported several advantages of the HIT design over its traditional c-Si counterpart, they are:

1. An intrinsic a-Si layer can act as an effective surface passivation layer for c-Si wafer.

2. The p+/n+ doped a-Si functions as an effective emitter/BSF for the cell.

3. The a-Si layers are deposited at much lower temperature, compared to the processing temperatures for traditional diffused c-Si technology.

4. The HIT cell has a lower temperature coefficient compared to c-Si cell technology.

Owing to all these advantages, this new hetero-junction solar cell is a considered to be a promising low cost alternative to traditional c-Si based solar cells.

Fabrication of HIT Cells

The details of the fabrication sequence vary from group to group. Typically, good quality, CZ/FZ grown c-Si wafer (with ~1ms lifetimes) are used as the absorber layer of HIT cells. Using alkaline etchants, such as, NaOH or $(CH_3)_4NOH$ the (100) surface of the wafer is textured to form the pyramids of 5-10µm height. Next, the wafer is cleaned using peroxide and HF solutions. This is followed by deposition of intrinsic a-Si passivation layer, typically through PECVD or Hot-wire CVD. The silane (SiH4) gas diluted with H_2 is used as a precursor. The deposition temperature and pressure is maintained at 200° C and 0.1-1 Torr. Precise control over this step is essential to avoid the formation of defective epitaxial Si. Cycles of deposition and annealing and H_2 plasma treatment are shown to have provided excellent surface passivation. Diborane or Trimethylboron gas mixed with SiH_4 is used to deposit p-type a-Si layer, while, Phosphine gas mixed with SiH_4 is used to deposit n-type a-Si layer. It should be noted that direct deposition of doped a-Si layers on c-Si wafer is shown to have very poor passivation properties. This is most likely due to dopant induced defect generation in a-Si layers. Sputtered indium tin oxide (ITO) is commonly used as a transparent conductive oxide (TCO) layer on top of the front and back a-Si layer in bi-facial design, as a-Si has high lateral resistance. It is generally deposited on the back side as well fully metallized cell to avoid diffusion of back metal and also for impedance matching for the reflected light. The silver/aluminum grid of 50-100μm thick is deposited through stencil printing for the front contact and back contact for bi-facial design.

Opto-electrical Modeling and Characterization of HIT Cells

The literature discusses several studies to interpret carrier transport bottlenecks in these cells. Traditional light and dark I-V are extensively studied and are observed to have several non-trivial features, which cannot be explained using the traditional solar cell diode theory. This is because of the presence of hetero-junction between the intrinsic a-Si layer and c-Si wafer which introduces additional complexities to current flow. In addition, there has been significant efforts to characterize this solar cell using C-V, impedance spectroscopy, surface photo-voltage, suns-Voc to produce complementary information.

Further, a number of design improvements, such as, the use of new emitters, bifacial configuration, interdigitated back contact (IBC) configuration bifacial-tandem configuration are actively being pursued.

Mono-silicon

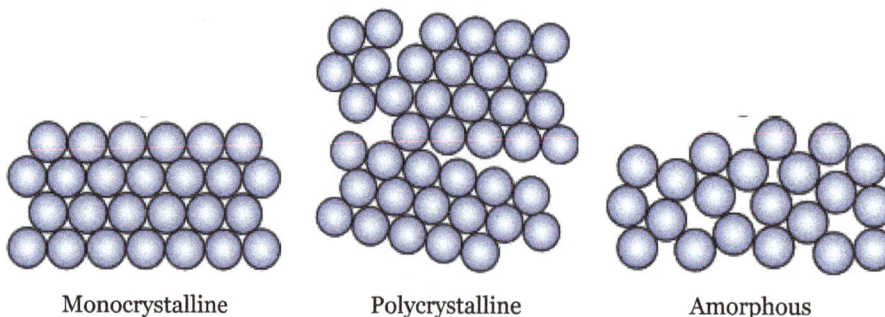

Monocrystalline   Polycrystalline   Amorphous

Schematic of allotropic forms of silicon

Monocrystalline silicon (mono c-Si) is a form in which the crystal structure is homogeneous throughout the material; the orientation, lattice parameter, and electronic properties are constant throughout the material. Dopant atoms such as phosphorus and boron are often incorporated into the film to make the silicon n-type or p-type respectively. Monocrystalline silicon is fabricated in the form of silicon wafers, usually by the Czochralski Growth method, and can be quite expensive depending on the radial size of the desired single crystal wafer (around $200 for a 300 mm Si wafer). This monocrystalline material, while useful, is one of the chief expenses associated with producing photovoltaics where approximately 40% of the final price of the product is attributable to the cost of the starting silicon wafer used in cell fabrication.

Multi-silicon

Multicrystalline silicon (multi c-Si) is composed of many smaller silicon grains of varied crystallographic orientation, typically >1 mm in size. This material can be synthesized easily by allowing liquid silicon to cool using a seed crystal of the desired crystal structure. Additionally, other methods for forming smaller-grained polycrystalline silicon (poly-Si) exist such as high temperature chemical vapor deposition (CVD).

Not Classified as Crystalline Silicon

These allotropic forms of silicon are not classified as crystalline silicon. They belong to the group of thin film solar cells.

Amorphous Silicon

Amorphous silicon (a-Si) has no long-range periodic order. The application of amorphous silicon to photovoltaics as a standalone material is somewhat limited by its inferior electronic properties. When paired with microcrystalline silicon in tandem and triple-junction solar cells, however, higher efficiency can be attained than with single-junction solar cells. This tandem assembly of solar cells allows one to obtain a thin-film material with a bandgap of around 1.12 eV (the same as single-crystal silicon) compared to the bandgap of amorphous silicon of 1.7-1.8 eV bandgap. Tandem solar cells are then attractive since they can be fabricated with a bandgap similar to single-crystal silicon but with the ease of amorphous silicon.

Nanocrystalline Silicon

Nanocrystalline silicon (nc-Si), sometimes also known as *microcrystalline silicon* (μc-Si), is a form of porous silicon. It is an allotropic form of silicon with paracrystalline structure—is similar to amorphous silicon (a-Si), in that it has an amorphous phase. Where they differ, however, is that nc-Si has small grains of crystalline silicon within the amorphous phase. This is in contrast to polycrystalline silicon (poly-Si) which consists solely of crystalline silicon grains, separated by grain boundaries. The difference comes solely from the grain size of the crystalline grains. Most materials with grains in the micrometre range are actually fine-grained polysilicon, so nanocrystalline silicon is a better term. The term Nanocrystalline silicon refers to a range of materials around the transition region from amorphous to microcrystalline phase in the silicon thin film.

Protocrystalline Silicon

Protocrystalline silicon has a higher efficiency than amorphous silicon (a-Si) and it has also been shown to improve stability, but not eliminate it. A protocrystalline phase is a distinct phase occurring during crystal growth which evolves into a microcrystalline form.

Protocrystalline Si also has a relatively low absorption near the band gap owing to its more ordered crystalline structure. Thus, protocrystalline and amorphous silicon can be combined in a tandem solar cell where the top layer of thin protocrystalline silicon absorbs short-wavelength light whereas the longer wavelengths are absorbed by the underlying a-Si substrate.

Transformation of Amorphous into Crystalline Silicon

Amorphous silicon can be transformed to crystalline silicon using well-understood and widely implemented high-temperature annealing processes. The typical method used in industry requires high-temperature compatible materials, such as special high temperature glass that is expensive to produce. However, there are many applications for which this is an inherently unattractive production method.

Low Temperature Induced Crystallization

Flexible solar cells have been a topic of interest for less conspicuous-integrated power generation than solar power farms. These modules may be placed in areas where traditional cells would not be feasible, such as wrapped around a telephone pole or cell phone tower. In this application, a photovoltaic material may be applied to a flexible substrate, often a polymer. Such substrates cannot survive the high temperatures experienced during traditional annealing. Instead, novel methods of crystallizing the silicon without disturbing the underlying substrate have been studied extensively. Aluminum-induced crystallization (AIC) and local laser crystallization are common in the literature, however not extensively used in industry.

In both of these methods, amorphous silicon is grown using traditional techniques such as plasma-enhanced chemical vapor deposition (PECVD). The crystallization methods diverge during post-deposition processing.

In aluminum-induced crystallization, a thin layer of aluminum (50nm or less) is deposited by physical vapor deposition onto the surface of the amorphous silicon. This stack of material is then annealed at a relatively low temperature between 140°C and 200°C in a vacuum. The aluminum that diffuses into the amorphous silicon is believed to weaken the hydrogen bonds present, allowing crystal nucleation and growth. Experiments have shown that polycrystalline silicon with grains on the order of 0.2 – 0.3µm can be produced at temperatures as low as 150°C. The volume fraction of the film that is crystallized is dependent on the length of the annealing process.

Aluminum-induced crystallization produces polycrystalline silicon with suitable crystallographic and electronic properties that make it a candidate for producing polycrystalline thin films for photovoltaics. AIC can be used to generate crystalline silicon nanowires and other nano-scale structures.

Another method of achieving the same result is the use of a laser to heat the silicon locally without heating the underlying substrate beyond some upper temperature limit. An excimer laser or, alternatively, green lasers such as a frequency-doubled Nd:YAG laser is used to heat the amorphous silicon, supplying energy necessary to nucleate grain growth. The laser fluence must be carefully controlled in order to induce crystallization without causing widespread melting. Crystallization of the film occurs as a very small portion of the silicon film is melted and allowed to cool. Ideally, the laser should melt the silicon film through its entire thickness, but not damage the substrate. Toward this end, a layer of silicon dioxide is sometimes added to act as a thermal barrier. This allows the use of substrates that cannot be exposed to the high temperatures of standard annealing, polymers for instance. Polymer-backed solar cells are of interest for seamlessly integrated power production schemes that involve placing photovoltaics on everyday surfaces.

A third method for crystallizing amorphous silicon is the use of thermal plasma jet. This strategy is an attempt to alleviate some of the problems associated with laser processing – namely the small region of crystallization and the high cost of the process on a production scale. The plasma torch is a simple piece of equipment that is used to thermally anneal the amorphous silicon. Compared to the laser method, this technique is simpler and more cost effective.

Plasma torch annealing is attractive because the process parameters and equipment dimension can be changed easily to yield varying levels of performance. A high level of crystallization (~90%) can be obtained with this method. Disadvantages include difficulty achieving uniformity in the crystallization of the film. While this method is applied frequently to silicon on a glass substrate, processing temperatures may be too high for polymers.

Polycrystalline Silicon

Polycrystalline silicon (Si), or polysilicon, refers to the raw material use in the production of single crystal wafers-the substrate for silicon-based solar cells and semiconductors. The raw material has a dark gray color, with a bronze to bluish metallic sheen. Meeting the demands of end-users for solar power systems begin with the manufacturing of high quality polycrystalline silicon, in short polysilicon.

Silicon-based solar panels make up 90 percent of the photovoltaic market. Thin-film technologies, which consist mainly of cadmium tellurium (CdTe) and copper indium gallium selenium (CIS/CIGS), cover the other ten percent. The thin film panel manufacturing process requires about 1/100th of the charging material needed for silicon-based modules. However, these panels have a lower efficient than standard solar modules.

Production of Polycrystalline Silicon Cells

Silica – Starting Raw Material

Silica rates second, behind oxygen, as the most abundant material on Earth. Nearly 26 percent of the Earth's crust consists of siliceous minerals, such as Chalcedony, Chert, Opal, Agate and Quartz, contain the element silicon, or a pure silica dioxide.

The starting point for hyperpure silicon begins with quartzite, a fine to median grain rock and a much sought after raw material for silicon. Quartzite has a whitish gray color and consists mostly of re-crystallized silica dioxide grains. It contains 90 to 99 percent quartz-- the highest concentration of pure silica on the planet.

Even so, the path from quartzite to hyperpure polycrystalline silicon involves an extensive, week-long process. To meet the stringent specifications required by the PV industry, the silicon undergoes a series of steps to process and refine the material into solar-grade silicon.

Refining Quartz to Trichlorosilane

Refinement begins in an electric arc furnace, which operates at a temperature of 1,800 to 2,000°Centigrade(C). The process uses COKE to reduce the quartzite to metallurgical-grade silicon (MG-Si). Although three different technologies exist for manufacturing polysilicon, most manufacturers employ the oldest process known as the Siemens method.

Hydrochlorination System: The MG-Si mixes with hydrogen gas and Hydrogen Chloride (HCL) at 350°C. Heating MG-Si in a fluidized bed reactor represents the first step in refining metallurgical-grade silicon to hyper pure multi crystalline silicon. This heated mixture produces trichlorosilane gas (TCS)-the starting material converted to polysilicon.

Hydrochlorination involves a closed-loop process, which means it recovers and recycles silicontetrachloride (STC) - a by-product from the polysilicon deposition process.

TCS Purification

The TCS undergoes an extensive distillation process, which requires four-steps to purify the material:

- Take out impurities, metal chlorides, and higher order chlorosilanes

- Separate and gather STC for conversion to TCS

- Remove hydrocarbons and light-boiling compounds

- Eliminate carbon chlorosilanes compounds

Hydrogen reduces the now high pure trichlorosilane, at around 1100°C, to make silicon again. Before the next step, the TCS is stored in towers to undergo a quality analysis to determine its approval for use.

On approval, the TCS flows through a pipeline to a chemical vapor deposition (CVD) reactor building to form silicon again. The next step in the process deposits the hyperpure silicon on electrically heated highly pure silicon rods. The elemental silicone grows on the rods uniformly in polycrystalline form.

Polycrystalline silicon production plant

This step forms rods of a specified diameter--around 130 millimeters. The rods—grown in batches, must then cool and transport to a clean area for further handling. This manufacturing process consumes huge amounts of energy. Some industry estimates put silicon production at 85 percent of the entire energy consumption required for manufacturing solar panels.

Manufacturers break hyperpure silicon rods into chunks, grade the material and package the material in industry standard containers with traceable lot numbers. Processing takes about one week. Polycrystalline manufacturers warehouse the product—later forwarding the product downstream to solar cell manufacturers.

Polycrystalline Silicon Purity

Although the PV market does not have as a stringent a requirement for purity as semiconductors, the PV industry does not tolerate impurities in excess of one parts per million (PPM) -- PV grade polysilicon ranges from 0.1 to 0.01 PPM-- a purity of 99.99999 to 99.999999 percent. The industry has a purity standard of 99.999999 or "six nines".

Polycrystalline silicon requires purity of only one foreign atom per 10 billion silicon atoms-- the equivalent of placing a penny on the area the size of 100 American-style football fields. In fact, many solar cell manufacturers have move to accepting only polysilicon of the highest purity, especially monocrystalline solar cell manufacturers.

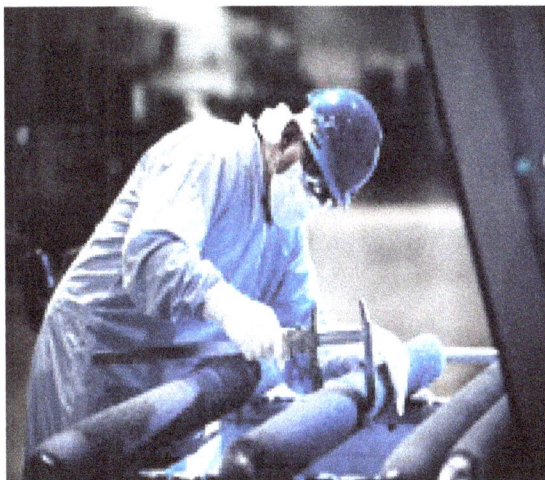

Worker checking polysilicon ingot specifications

Deposition Methods

Polysilicon deposition, or the process of depositing a layer of polycrystalline silicon on a semiconductor wafer, is achieved by the chemical decomposition of silane (SiH_4) at high temperatures of 580 to 650°C. This pyrolysis process releases hydrogen.

$$SiH_4(g) \rightarrow Si(s) + 2\ H_2(g) \text{ CVD at 500-800°C}$$

Polysilicon layers can be deposited using 100% silane at a pressure of 25–130 Pa (0.19–0.98 Torr) or with 20–30% silane (diluted in nitrogen) at the same total pressure. Both of these processes can deposit polysilicon on 10–200 wafers per run, at a rate of 10–20 nm/min and with thickness uniformities of ±5%. Critical process variables for polysilicon deposition include temperature, pressure, silane concentration, and dopant concentration. Wafer spacing and load size have been shown to have only minor effects on the deposition process. The rate of polysilicon deposition increases rapidly with temperature, since it follows Arrhenius behavior, that is deposition rate = $A \cdot \exp(-qE_a/kT)$ where q is electron charge and k is the Boltzmann constant. The activation energy (E_a) for polysilicon deposition is about 1.7 eV. Based on this equation, the rate of polysilicon deposition increases as the deposition temperature increases. There will be a minimum temperature, however, wherein the rate of deposition becomes faster than the rate at which unreacted silane arrives at the surface. Beyond this temperature, the deposition rate can no longer increase with temperature, since it is now being hampered by lack of silane from which the polysilicon will be generated. Such a reaction is then said to be 'mass-transport-limited.' When a polysilicon deposition process becomes mass-transport-limited, the reaction rate becomes dependent primarily on reactant concentration, reactor geometry, and gas flow.

When the rate at which polysilicon deposition occurs is slower than the rate at which unreacted silane arrives, then it is said to be surface-reaction-limited. A deposition process that is surface-reaction-limited is primarily dependent on reactant concentration and reaction temperature. Deposition processes must be surface-reaction-limited because they result in excellent thickness uniformity and step coverage. A plot of the logarithm of the deposition rate against the reciprocal of the absolute temperature in the surface-reaction-limited region results in a straight line whose slope is equal to $-qE_a/k$.

At reduced pressure levels for VLSI manufacturing, polysilicon deposition rate below 575°C is too slow to be practical. Above 650°C, poor deposition uniformity and excessive roughness will be encountered due to unwanted gas-phase reactions and silane depletion. Pressure can be varied inside a low-pressure reactor either by changing the pumping speed or changing the inlet gas flow into the reactor. If the inlet gas is composed of both silane and nitrogen, the inlet gas flow, and hence the reactor pressure, may be varied either by changing the nitrogen flow at constant silane flow, or changing both the nitrogen and silane flow to change the total gas flow while keeping the gas ratio constant. Recent investigations have shown that e-beam evaporation, followed by SPC (if needed) can be a cost effective and faster alternative for producing solar grade poly-Si thin films. Modules produced by such method are shown to have a photovoltaic efficiency of ~6%.

Polysilicon doping, if needed, is also done during the deposition process, usually by adding phosphine, arsine, or diborane. Adding phosphine or arsine results in slower deposition, while adding diborane increases the deposition rate. The deposition thickness uniformity usually degrades when dopants are added during deposition.

Upgraded Metallurgical-grade Silicon

Upgraded metallurgical-grade (UMG) silicon (also known as UMG-Si) solar cell is being produced as a low cost alternative to polysilicon created by the Siemens process. UMG-Si greatly reduces impurities in a variety of ways that require less equipment and energy than the Siemens process. It is about 99% pure which is three or more orders of magnitude less pure and about 10 times less expensive than polysilicon ($1.70 to $3.20 per kg from 2005 to 2008 compared to $40 to $400 per kg for polysilicon). It has the potential to provide nearly-as-good solar cell efficiency at 1/5 the capital expenditure, half the energy requirements, and less than $15/kg.

Siemens process   FBR process

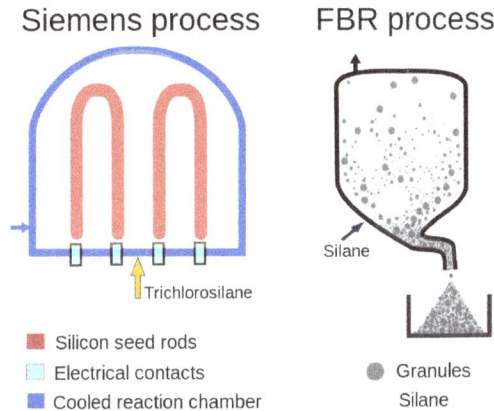

Trichlorosilane

■ Silicon seed rods
□ Electrical contacts
■ Cooled reaction chamber

Silane

● Granules
Silane

Schematic diagram of the traditional Siemens and the Fluidized bed reactor purification process.

In 2008 several companies were touting the potential of UMG-Si in 2010, but the credit crisis greatly lowered the cost of polysilicon and several UMG-Si producers put plans on hold. The Siemens process will remain the dominant form of production for years to come due to more efficiently implementing the Siemens process. GT Solar claims a new Siemens process can produce at $27/kg and may reach $20/kg in 5 years. GCL-Poly expects production costs to be $20/kg by end of 2011. Elkem Solar estimates their UMG costs to be $25/kg, with a capacity of 6,000 tonnes by the end of 2010. Calisolar expects UMG technology to produce at $12/kg in 5 years with boron at 0.3 ppm and phosphorus at 0.6 ppm. At $50/kg and 7.5 g/W, module manufacturers spend $0.37/W for the polysilicon. For comparison, if a CdTe manufacturer pays spot price for tellurium ($420/kg in April 2010) and has a 3 μm thickness, their cost would be 10 times less, $0.037/Watt. At 0.1 g/W and $31/ozt for silver, polysilicon solar producers spend $0.10/W on silver.

Q-Cells, Canadian Solar, and Calisolar have used Timminco UMG. Timminco is able to produce UMG-Si with 0.5 ppm boron for $21/kg but were sued by shareholders because they had expected $10/kg. RSI and Dow Corning have also been in litigation over UMG-Si technology.

Potential for use of Polycrystalline Silicon

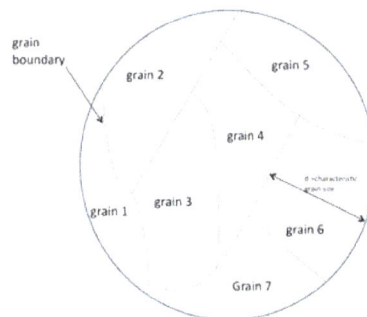

A picture of grain boundaries for polysilicon

Each grain is crystalline over the width of the grain. The grain boundary separates the grains where the adjoining grain is at a different orientation from its neighbor. The grain boundary separates regions of different crystal structure thus serving as a center for recombination. 'd' here is a characteristic grain size, which should be maximized for maximum solar cell efficiency. Typical values of d are about 1 micrometre.

Currently, polysilicon is commonly used for the conducting gate materials in semiconductor devices such as MOSFETs; however, it has potential for large-scale photovoltaic devices. The abundance, stability, and low toxicity of silicon, combined with the low cost of polysilicon relative to single crystals makes this variety of material attractive for photovoltaic production. Grain size has been shown to have an effect on the efficiency of polycrystalline solar cells. Solar cell efficiency increases with grain size. This effect is due to reduced recombination in the solar cell. Recombination, which is a limiting factor for current in a solar cell, occurs more prevalently at grain boundaries.

The resistivity, mobility, and free-carrier concentration in monocrystalline silicon vary with doping concentration of the single crystal silicon. Whereas the doping of polycrystalline silicon does have an effect on the resistivity, mobility, and free-carrier concentration, these properties strongly depend on the polycrystalline grain size, which is a physical parameter that the material scientist can manipulate. Through the methods of crystallization to form polycrystalline silicon, an engineer can control the size of the polycrystalline grains which will vary the physical properties of the material.

Novel Ideas for Polycrystalline Silicon

The use of polycrystalline silicon in the production of solar cells requires less material and therefore provides higher profits and increased manufacturing throughput. Polycrystalline silicon does not need to be deposited on a silicon wafer to form a solar cell, rather it can be deposited on other-cheaper materials, thus reducing the cost. Not requiring a silicon wafer alleviates the silicon shortages occasionally faced by the microelectronics industry. An example of not using a silicon wafer is crystalline silicon on glass (CSG) materials

A primary concern in the photovoltaics industry is cell efficiency. However, sufficient cost savings from cell manufacturing can be suitable to offset reduced efficiency in the field, such as the use of larger solar cell arrays compared with more compact/higher efficiency designs. Designs such as CSG are attractive because of a low cost of production even with reduced efficiency. Higher efficiency devices yield modules that occupy less space and are more compact; however, the 5–10% efficiency of typical CSG devices still makes them attractive for installation in large central-service stations, such as a power station. The issue of efficiency versus cost is a value decision of whether one requires an "energy dense" solar cell or sufficient area is available for the installation of less expensive alternatives. For instance, a solar cell used for power generation in a remote location might require a more highly efficient solar cell than one used for low-power applications, such as solar accent lighting or pocket calculators, or near established power grids.

Advantages

- The process used to make polycrystalline silicon is simpler and cost less. The amount of waste silicon is less compared to monocrystalline.

- Polycrystalline solar panels tend to have slightly lower heat tolerance than monocrystalline solar panels. This technically means that they perform slightly worse than monocrystalline solar panels in high temperatures. Heat can affect the performance of solar panels and shorten their lifespans. However, this effect is minor, and most homeowners do not need to take it into account.

Disadvantages

- The efficiency of polycrystalline-based solar panels is typically 13-16%. Because of lower silicon purity, polycrystalline solar panels are not quite as efficient as monocrystalline solar panels.

- Lower space-efficiency: You generally need to cover a larger surface to output the same electrical power as you would with a solar panel made of monocrystalline silicon. However, this does not mean every monocrystalline solar panel perform better than those based on polycrystalline silicon.

- Monocrystalline and thin-film solar panels tend to be more aesthetically pleasing since they have a more uniform look compared to the speckled blue color of polycrystalline silicon.

Cadmium Telluride

Cadmium telluride (CdTe) is a photovoltaic (PV) technology based on the use of a thin film of CdTe to absorb and convert sunlight into electricity. CdTe is growing rapidly in acceptance and now represents the second most utilized solar cell material in the world. The first is still silicon.

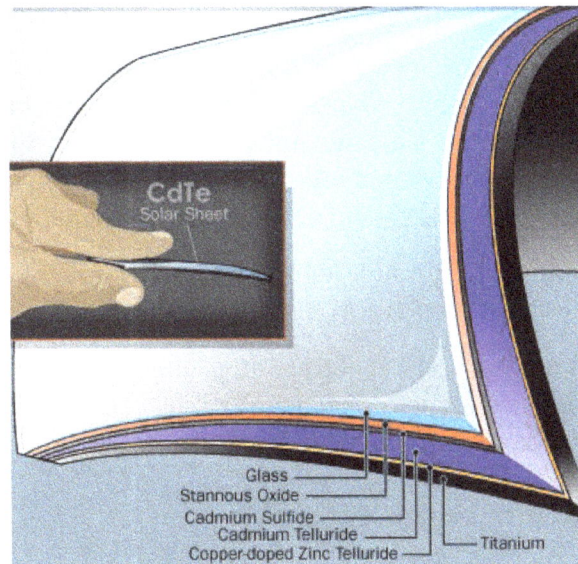

How thin-film solar cells work

Solar panels based on CdTe are the first and only thin film photovoltaic technology to surpass crystalline silicon PV in cheapness for a significant portion of the PV market, namely in multi-kilowatt systems.

Cell Efficiency

Best cell efficiency has plateaued at 16.5% since 2001 (a record held by NREL). The opportunity to increase current has been almost fully exploited, but more difficult challenges associated with junction quality, with properties of CdTe and with contacting have not been as successful.

Improved doping of CdTe and increased understanding of key processing steps (e.g., cadmium chloride recrystallization and contacting) are key to improving cell efficiency. Since CdTe has the optimal band gap for single-junction devices, it may be expected that efficiencies close to exceeding 20% (such as already shown in CIS alloys) should be achievable in mass produced CdTe cells.

In 2009, EMPA, the Swiss Federal Laboratories for Materials Testing and Research, demonstrated a 12.4% efficient solar cell on flexible plastic substrate.

Low Cost Manufacturing

The major advantage of this technology is that the panels can manufactured at lower costs than silicon based solar panels. First Solar was the first manufacturer of Cadmium telluride panels to produced solar cells for less than $1.00 per watt.

Some experts believe it will be possible to get the solar cell costs down to around $0.5 per watt. With commodity-like margins and combined with balance-of-system (BOS) costs, installed systems near $1.5/W seem achievable. With sufficient levels of sunlight – this would allow such systems to produce electricity in the $0.06 to $0.08 / kWh range – or for less than fuel based electricity costs.

Advantages of Cadmium Telluride Solar Panels

CdTe panels have several advantages over traditional silicon technology. These include:

- Ease of manufacturing: The necessary electric field, which makes turning solar energy into electricity possible, stems from properties of two types of cadmium molecules, cadmium sulfide and cadmium telluride. This means a simple mixture of molecules achieves the required properties, simplifying manufacturing compared to the multi-step process of joining two different types of doped silicon in a silicon solar panel.

- Good match with sunlight: Cadmium telluride absorbs sunlight at close to the ideal wavelength, capturing energy at shorter wavelengths than is possible with silicon panels.

- Cadmium is abundant: Cadmium is abundant, produced as a by-product of other important industrial metals such as zinc, consequently it has not had the wider price swings that have happened in the past two years with silicon prices.

Drawbacks of Cadmium Telluride

While price is a major advantage, there are some drawbacks to this type of solar panels, namely:

1. Lower efficiency levels

   Cadmium telluride solar panels currently achieve an efficiency of 10.6%, which is significantly lower than the typical efficiencies of silicon solar cells.

2. Tellurium supply

   While Cadmium is relatively abundant, Tellurium is not. Tellurium (Te) is an extremely rare element (1-5 parts per billion in the Earth's crust. According to USGS, global tellurium production in 2007 was 135 metric tons. Most of it comes as a by-product of copper, with smaller byproduct amounts from lead and gold. One gigawatt (GW) of CdTe PV modules would require about 93 metric tons (at current efficiencies and thicknesses), so the availability of tellurium will eventually limited how many panels can be produced with this material.

Since CdTe is now regarded as an important technology in terms of PV's future impact on global energy and environment, the issue of tellurium availability is significant. Recently, researchers have added an unusual twist – astrophysicists identify tellurium as the most abundant element in the universe with an atomic number over 40. This surpasses, e.g., heavier materials like tin, bismuth, and lead, which are common. Researchers have shown that well-known undersea ridges (which are now being evaluated for their economic recoverability) are rich in tellurium and by themselves could supply more tellurium than we could ever use for all of our global energy. It is not yet known whether this undersea tellurium is recoverable, nor whether there is much more tellurium elsewhere that can be recovered.

Capacity for manufacturing thin-film photovoltaic solar cells from cadmium telluride is very close to the maximum supply of tellurium available, or that may become available and that the ability of companies like First Solar to continue to expand at the rates they have been growing at over the past several years will become increasingly difficult to maintain because of lack of available tellurium (even with recovery from recyclying).

3. Toxicity of Cadmium

Cadmium is one of the top 6 deadliest and toxic materials known. However, CdTe appears to be less toxic than elemental cadmium, at least in terms of acute exposure.

This is not to say it is harmless. Cadmium telluride is toxic if ingested, if its dust is inhaled, or if it is handled improperly (i.e. without appropriate gloves and other safety precautions). The toxicity is not solely due to the cadmium content. One study found that the highly reactive surface of cadmium telluride quantum dots triggers extensive reactive oxygen damage to the cell membrane, mitochondria, and cell nucleus. In addition, the cadmium telluride films are typically recrystallized in a toxic compound of cadmium chloride.

The disposal and long term safety of cadmium telluride is a known issue in the large-scale commercialization of cadmium telluride solar panels. Serious efforts have been made to understand and overcome these issues. Researchers from the U.S. Department of Energy's Brookhaven National Laboratory have found that large-scale use of CdTe PV modules does not present any risks to health and the environment, and recycling the modules at the end of their useful life resolves any environmental concerns. During their operation, these modules do not produce any pollutants, and furthermore, by displacing fossil fuels, they offer great environmental benefits. CdTe PV modules appear to be more environmentally friendly than all other current uses of Cd.

The approach to CdTe safety in the European Union and China is however, much more cautious: cadmium and cadmium compounds are considered as toxic carcinogens in EU whereas China regulations allow Cd products for export only. The issue about regulating the use of Cadmium Telluride is currently being discussed in Europe.

At the present time – the most common opinion is that the use of Cadmium Terlluride in residential / industrial rooftop installations does not pose a major environmental problem.

Amorphous Silicon

Amorphous silicon (a-Si) is the non-crystalline form of silicon. It is the most well developed of the thin film technologies having been on the market for more than 15 years. It is widely used in pocket calculators, but it also powers some private homes, buildings, and remote facilities.

United Solar Systems Corp. (UniSolar) pioneered amorphous-silicon solar cells and remains a major maker today, as does Sharp and Sanyo.

Manufacturing

Amorphous silicon panels are formed by vapor-depositing a thin layer of silicon material – about 1 micrometer thick – on a substrate material such as glass or metal. Amorphous silicon can also be deposited at very low temperatures, as low as 75 degrees Celsius, which allows for deposition on plastic as well.

In its simplest form, the cell structure has a single sequence of p-i-n layers. However, single layer cells suffer from significant degradation in their power output (in the range 15-35%) when exposed to the sun. The mechanism of degradation is called the Staebler-Wronski Effect, after its discoverers.

Better stability requires the use of a thinner layers in order to increase the electric field strength across the material. However, this reduces light absorption, hence cell efficiency. This has led the industry to develop tandem and even triple layer devices that contain p-i-n cells stacked one on top of the other.

One of the pioneers of developing solar cells using amorphous silicon is Uni-Solar. They use a triple layer system that is optimized to capture light from the full solar spectrum).

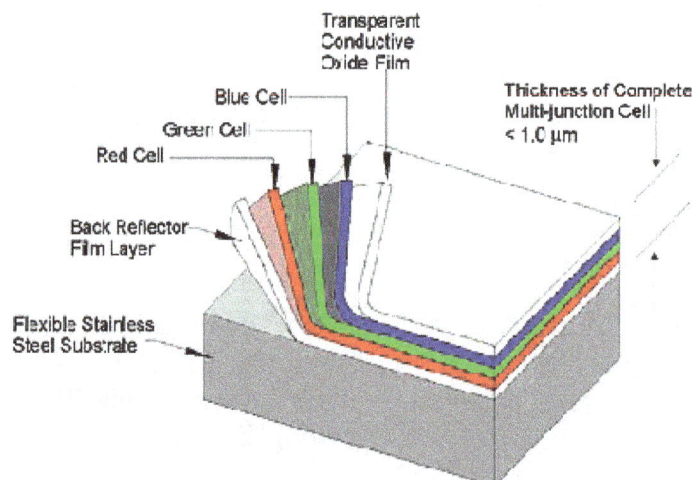

The thickness of the solar cell is just 1 micron, or about 1/300th the size of mono-crystalline silicon solar cell.

Efficiency

While crystalline silicon achieves a yield of about 18 percent, amorphous solar cells' yield remains at around 7 percent. The low efficiency rate is partly due to the Staebler-Wronski effect, which manifests itself in the first hours when the panels are exposed to sunlight, and results in a decrease in the energy yield of an amorphous silicon panel from 10 percent to around 7 percent.

A German researcher has demonstrated how to raise the energy output of amorphous silicon solar panels from around 7 percent to 9 percent, Gijs van Elzakker investigated adaptations in the production processes of amorphous silicon modules to increase the output without any additional costs using Silane Gas to reduce the Staebler-Wronski effect.

This is just one approach being tried today. UniSolar's, laminate efficiency is currently at 8.2%; however, by late spring 2011, the company expects to be at 10% using their triple coating / triple junction technology.

Banking on improvements in light-trapping, high-rate deposition, and a HybridNano technology, Uni-Solar expects to be able to push it's conversion efficiency to 12% by 2012 and believes it has the potential to reach 20+% for its product line.

Amorphous Silicon and Carbon

Amorphous alloys of silicon and carbon (amorphous silicon carbide, also hydrogenated, $a\text{-}Si_{1-x}C_x\text{:}H$) are an interesting variant. Introduction of carbon atoms adds extra degrees of freedom for control of the properties of the material. The film could also be made transparent to visible light.

Increasing the concentration of carbon in the alloy widens the electronic gap between conduction and valence bands (also called "optical gap" and bandgap). This can potentially increase the light efficiency of solar cells made with amorphous silicon carbide layers. On the other hand, the electronic properties as a semiconductor (mainly electron mobility), are adversely affected by the increasing content of carbon in the alloy, due to the increased disorder in the atomic network.

Several studies are found in the scientific literature, mainly investigating the effects of deposition parameters on electronic quality, but practical applications of amorphous silicon carbide in commercial devices are still lacking.

Properties

The density of amorphous Si has been calculated as 4.90×10^{22} atom/cm^3 (2.285 g/cm^3) at 300 K. This was done using thin (5 micron) strips of amorphous silicon. This density is 1.8±0.1% less dense than crystalline Si at 300 K. Silicon is one of the few elements that expands upon cooling and has a lower density as a solid than as a liquid.

Hydrogenated Amorphous Silicon

Unhydrogenated a-Si has a very high defect density which leads to undesirable semiconductor properties such as poor photoconductivity and prevents doping which is critical to engineering semiconductor properties. By introducing hydrogen during the fabrication of amorphous silicon, photoconductivity is significantly improved and doping is made possible. Hydrogenated amorphous silicon, a-Si:H, was first fabricated in 1969 by Chittick, Alexander and Sterling by deposition using a silane gas (SiH_4) precursor. The resulting material showed a lower defect density and increased conductivity due to impurities. Interest in a-Si:H came when (in 1975), LeComber and Spear discovered the ability for substitutional doping of a-Si:H using phosphine (n-type) or diborane (p-type). The role of hydrogen in reducing defects was verified by Paul's group at Harvard who found a hydrogen concentration of about 10 atomic % through IR vibration, which for Si-H bonds has a frequency of about 2000 cm^{-1}. Starting in the 1970s, a-Si:H was developed in solar cells by RCA by which steadily climbed in efficiency to about 13.6% in 2015.

Deposition Processes

	CVD	PECVD	Catalytic CVD	Sputtering
Type of film	a-Si:H	a-Si:H	a-Si:H	a-Si
Unique application		Large-area electronics		Hydrogen-free deposition
Chamber temperature	600°C	30–300°C		30–1000°C
Active element temperature			2000°C	
Chamber pressure	0.1–10 Torr	0.1–10 Torr		0.001–0.1 Torr
Physical principle	Thermolysis	Plasma-induced dissociation	Thermolysis	Ionization of Si source
Facilitators			W/Ta heated wires	Argon cations
Typical drive voltage		RF 13.56 MHz; 0.01-1W/cm²		
Si source	SiH_4 gas	SiH_4 gas	SiH_4 gas	crucible
Substrate temperature	controllable	controllable	controllable	controllable

Applications

While a-Si suffers from lower electronic performance compared to c-Si, it is much more flexible in its applications. For example, a-Si layers can be made thinner than c-Si, which may produce savings on silicon material cost.

One further advantage is that a-Si can be deposited at very low temperatures, e.g., as low as 75 degrees Celsius. This allows deposition on not only glass, but plastic as well, making it a candidate for a roll-to-roll processing technique. Once deposited, a-Si can be doped in a fashion similar to c-Si, to form p-type or n-type layers and ultimately to form electronic devices.

Another advantage is that a-Si can be deposited over large areas by PECVD. The design of the PECVD system has great impact on the production cost of such panel, therefore most equipment suppliers put their focus on the design of PECVD for higher throughput, that leads to lower manufacturing cost particularly when the silane is recycled.

Arrays of small (under 1 mm by 1 mm) a-Si photodiodes on glass are used as visible-light image sensors in some flat panel detectors for fluoroscopy and radiography.

Photovoltaics

The "Teal Photon" solar-powered calculator produced in the late 1970s

Amorphous silicon (a-Si) has been used as a photovoltaic solar cell material for devices which require very little power, such as pocket calculators, because their lower performance compared to conventional crystalline silicon (c-Si) solar cells is more than offset by their simplified and lower cost of deposition onto a substrate. The first solar-powered calculators were already available in the late 1970s, such as the Royal *Solar 1*, Sharp *EL-8026*, and Teal *Photon*.

More recently, improvements in a-Si construction techniques have made them more attractive for large-area solar cell use as well. Here their lower inherent efficiency is made up, at least partially, by their thinness – higher efficiencies can be reached by stacking several thin-film cells on top of each other, each one tuned to work well at a specific frequency of light. This approach is not applicable to c-Si cells, which are thick as a result of its indirect band-gap and are therefore largely opaque, blocking light from reaching other layers in a stack.

The source of the low efficiency of amorphous silicon photovoltaics is due largely to the low hole mobility of the material. This low hole mobility has been attributed to many physical aspects of the material, including the presence of dangling bonds (silicon with 3 bonds), floating bonds (silicon with 5 bonds), as well as bond reconfigurations. While much work has been done to control these sources of low mobility, evidence suggests that the multitude of interacting defects may lead to the mobility being inherently limited, as reducing one type of defect leads to formation others.

The main advantage of a-Si in large scale production is not efficiency, but cost. a-Si cells use only a fraction of the silicon needed for typical c-Si cells, and the cost of the silicon has historically been a significant contributor to cell cost. However, the higher costs of manufacture due to the multi-layer construction have, to date, made a-Si unattractive except in roles where their thinness or flexibility are an advantage.

Typically, amorphous silicon thin-film cells use a p-i-n structure. The placement of the p-type layer on top is also due to the lower hole mobility, allowing the holes to traverse a shorter average distance for collection to the top contact. Typical panel structure includes front side glass, TCO, thin-film silicon, back contact, polyvinyl butyral (PVB) and back side glass. Uni-Solar, a division of Energy Conversion Devices produced a version of flexible backings, used in roll-on roofing products. However, the world's largest manufacturer of amorphous silicon photovoltaics had to file for bankruptcy in 2012, as it could not compete with the rapidly declining prices of conventional solar panels.

Microcrystalline and Micromorphous Silicon

Microcrystalline silicon (also called nanocrystalline silicon) is amorphous silicon, but also contains small crystals. It absorbs a broader spectrum of light and is flexible. Micromorphous silicon module technology combines two different types of silicon, amorphous and microcrystalline silicon, in a top and a bottom photovoltaic cell. Sharp produces cells using this system in order to more efficiently capture blue light, increasing the efficiency of the cells during the time where there is no direct sunlight falling on them. Protocrystalline silicon is often used to optimize the open circuit voltage of a-Si photovoltaics.

Large-scale Production

Xunlight Corporation, which has received over $40 million of institutional investments, has completed the installation of its first 25 MW wide-web, roll-to-roll photovoltaic manufacturing equipment for the production of thin-film silicon PV modules. Anwell Technologies has also completed the installation of its first 40 MW a-Si thin film solar panel manufacturing facility in Henan with its in-house designed multi-substrate-multi-chamber PECVD equipment.

Photovoltaic Thermal Hybrid Solar Collectors

Photovoltaic thermal hybrid solar collectors (PVT) are systems that convert solar radiation into electrical energy and thermal energy. These systems combine a solar cell, which converts electromagnetic radiation (photons) into electricity, with a solar thermal collector, which captures the remaining energy and removes waste heat from the solar PV module. Solar cells suffer from a drop in efficiency with the rise in temperature due to increased resistance. Most such systems can be engineered to carry heat away from the solar cells thereby cooling the cells and thus improving their efficiency by lowering resistance. Although this is an effective method, it causes the thermal component to under-perform compared to a solar thermal collector. Recent research showed that a-Si:H PV with low temperature coefficients allow the PVT to be operated at high temperatures, creating a more symbiotic PVT system and improving performance of the a-Si:H PV by about 10%.

Thin-film-transistor Liquid-crystal Display

Amorphous silicon has become the material of choice for the active layer in thin-film transistors (TFTs), which are most widely used in large-area electronics applications, mainly for liquid-crystal displays (LCDs).

Thin-film-transistor liquid-crystal display (TFT-LCD) show a similar circuit layout process to that of semiconductor products. However, rather than fabricating the transistors from silicon, that is formed into a crystalline silicon wafer, they are made from a thin film of amorphous silicon that is deposited on a glass panel. The silicon layer for TFT-LCDs is typically deposited using the PECVD process. Transistors take up only a small fraction of the area of each pixel and the rest of the silicon film is etched away to allow light to easily pass through it.

Polycrystalline silicon is sometimes used in displays requiring higher TFT performance. Examples include small high-resolution displays such as those found in projectors or viewfinders. Amorphous silicon-based TFTs are by far the most common, due to their lower production cost, whereas polycrystalline silicon TFTs are more costly and much more difficult to produce.

Advantages

The principal advantage of amorphous silicon solar cells is their lower manufacturing costs, which makes these cells very cost competitive.

One of the main advantages of a-Si over crystalline silicon is that it is much more uniform over large areas. Since amorphous silicon is full of defects naturally, any other defects, such as impurities, do not affect the overall characteristics of the material too drastically.

Amporphous silicon can be produced in a variety of shapes and sizes (e.g., round, square, hexagonal, or any other complex shape. This makes it an ideal technology to use in a variety of applications such as powering electronic calculators, solar wristwatches, garden lights, and to power car accessories. Small solar cells used in pocket calculators have been made with a-Si for many years.

Unlike crystalline solar cells in which cells are cut apart and the recombined, amorphous silicon cells can be connected in series at the same time the cells are formed, making it is easy to create panels in a variety of voltages (e.g, for use in solar battery rechargers).

The human eye is sensitive to light with wavelengths of 400 nm to 700 nm. Since amorphous silicon solar cells are sensitive to light with essentially the same wavelengths, this means that in addition to be used as solar cells they can also be used as light sensors (e.g., outdoor sensor lights, etc).

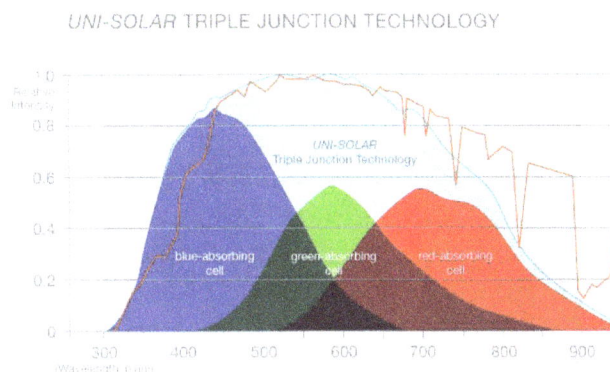

UNI-SOLAR TRIPLE JUNCTION TECHNOLOGY

Some amorphous solar panels also come with shade-resistant technology or multiple circuits within the cells, so that if an entire row of cells is subject to complete shading, the circuit won't be completely broken and some output can still be gained. This is especially useful when installing solar panels on a boat.

The development process of a-Si solar panels also renders them much less susceptible to breakage during transport or installation. This can help reduce the risk of damaging your significant investment in a photovoltaic system.

Another principal advantage of this type of technology is greater resistance to heat. According to a four year NREL study – it was observed that amorphous silicon PV modules experience higher results as temperatures increase.

Disadvantages

As mentioned previously, these panels have a lower efficiency than mono-crystalline solar cells, or even poly-crystalline solar cells. Attempts to increase the efficiency, such as building multi-layer cells or alloying with germanium to reduce its band gap and further improve light absorption all have an added complexity. Namely, the processes are more complex and process yields are likely to be lower and costs are likely to be higher as a result – thus reducing the cost advantage of this type of solar cell.

The expected lifetime of amorphous cells is shorter than the lifetime of crystalline cells, although how much shorter is difficult to determine, especially as the technology continues to evolve. From reading through the literature, it appears that the expected life is still in the order of 25 years or so. For example, Uni-Solar offers the following performance guarantee on their 144 Wp panels: 92% at 10 years, 84% at 20 years , 80% at 25 year (of minimum power).

Gallium Nitride

Gallium nitride (GaN) is a wide band gap semiconductor material and is the most popular material after silicon in the semiconductor industry. The prime movers behind this trend are LEDs, microwave, and more recently, power electronics. New areas of research also include Spintronics and Nano ribbon transistors, which leverage some of the unique properties of GaN. GaN has electron mobility comparable with silicon, but with a band gap that is three times larger, making it an excellent candidate for highpower applications and high-temperature operation. The ability to form thin AlGaN/GaN hetero structures, which exhibit the 2-D electron gas phenomenon, leads to high-electron mobility transistors, which exhibit high Johnson's figure of merit. Another interesting direction for GaN research, which is largely unexplored, is GaN based micromechanical devices or GaN micro electromechanical systems (MEMS). To fully unlock the potential of GaN and realize new advanced all GaN integrated circuits, it is essential to integrate passive devices (such as resonators and filters), sensors (such as temperature and gas sensors), and other more than Moore functional devices with GaN active electronics. Therefore, there is a growing interest in the use of GaN as a mechanical material.

Gallium Nitride: There has been increasing research and industrial activity in the area of gallium nitride (GaN) electronics, stimulated first by the successful demonstration of GaN LEDs. This set the field of GaN electronics in motion, and today the technology is improving the performance of several applications including use in transistors and using them as semiconductor material in solar photovoltaic. These may provide up to 5 times better than the silicon carbide products Gallium

Nitride is direct band gap semiconductor having Wurtzite crystal structure.

Formation: GaN crystals can be grown from a molten Ga melt held under 100 atmospheres of pressure of N_2 at 750°C. As Ga will not react with N2 below 1000°C, the powder must be made from something more reactive, usually in one of the following ways:

$$2Ga + 2NH_3 \rightarrow 2GaN + 3H_2$$

$$Ga_2O_3 + 2NH_3 \rightarrow 2GaN + 3H_2O$$

GaN Transistors

GaN transistors have been developed so as to meet the future demands of electricity. These are also compressed up into compact, lightweight design along with the improvement in the efficiency of the photovoltaic cells. These may provide up to 5 times better than the silicon carbide products and by enabling the inverter to be reduced to a significantly more compact package. The 650V GaN transistors played an indispensable role in obtaining the desired operating ranges of plates. The system stability may also be enhanced across a wide range of power levels, aiding the higher level of inverter integration. It uses future-forward design paired with a disruptive level of system integration. With this approach, we may reduce the installed cost of solar power with battery storage by up to 50% .Using an intelligent design capable of compensating for diverse power environments, these photovoltaic cells have been integrated creating a new category of solar photovoltaic cells. These have builtin battery storages that can provide electricity with electrically -optimized power efficiency. These types of solar plates have also been integrated with the GaN Systems' gallium nitride (GaN) transistors in its newly developed inverters to increase power efficiency and reduce size and weight within the enclosed solar plates. Also for the additional power requirements, standard IEC power outlet may be used.

References

- S. A. Campbell (2001), The Science and Engineering of Microelectronic Fabrication (2nd ed.), New York: Oxford University Press, ISBN 0-19-513605-5

- Cadmium-telluride: solar-facts-and-advice.com, Retrieved 11 May 2018

- Kishore, R.; Hotz, C.; Naseem, H. A. & Brown, W. D. (2001), "Aluminum-Induced Crystallization of Amorphous Silicon (α-Si:H) at 150°C", Electrochemical and Solid State Letters, 4 (2): G14–G16, doi:10.1149/1.1342182.

- Solar-photovoltaic-cell-basics: energy.gov, Retrieved 18 June 2018

- "Personal finance news, articles, tips and advice on managing your money - myfinances.co.uk" (PDF). My Finances. Retrieved 10 April 2018.

- Ghosh, Amal K.; Fishman, Charles & Feng, Tom (1980), "Theory of the electrical and photovoltaic properties of polycrystalline silicon", Journal of Applied Physics, 51 (1): 446, Bibcode:1980JAP....51..446G, doi:10.1063/1.327342.

- Crystalline-silicon-photovoltaics-research, solar: energy.gov, Retrieved 31 March 2018

- Morgan, D. V.; Board, K. (1991). An Introduction To Semiconductor Microtechnology (2nd ed.). Chichester, West Sussex, England: John Wiley & Sons. p. 27. ISBN 0471924784.

- Shah, A. V.; et al. (2003), "Material and solar cell research in microcrystalline silicon", Solar Energy Materials and Solar Cells, 78 (1–4): 469–491, doi:10.1016/S0927-0248(02)00448-8.

- Polysilicon-Manufacturing-20111117: energytrend.com, Retrieved 21 April 2018

- "Solar Polysilicon Manufacturers Cranking Out Supply Despite Losses - Solar Industry". solarindustrymag. com. 2 October 2012. Retrieved 10 April 2018.

- Banerjee, A.; Guha, S. (1991-01-15). "Study of back reflectors for amorphous silicon alloy solar cell application". Journal of Applied Physics. 69 (2): 1030–1035. Bibcode:1991JAP....69.1030B. doi:10.1063/1.347418. ISSN 0021-8979.

- Best-solar-panel-monocrystalline-polycrystalline-thin-film, polycrystalline: energyinformative.org, Retrieved 19 May 2018

3

Photovoltaic System and its Components

Generally, a photovoltaic system is a combination of solar panels comprising of a number of solar cells. It can be ground-mounted, wall-mounted or rooftop-mounted. The mount may either be fixed or a solar tracker affixed to follow the sun across the sky. This chapter has been carefully written to provide an understanding of photovoltaic systems and their components such as photovoltaic cell, solar panel, solar shingle, solar cable, etc.

Photovoltaic System

Figure: A photovoltaic system comprised of a solar panel array, inverter and other electrical hardware.

A photovoltaic (PV) system is a system composed of one or more solar panels combined with an inverter and other electrical and mechanical hardware that use energy from the Sun to generate electricity. PV systems can vary greatly in size from small rooftop or portable systems to massive utility-scale generation plants. Although PV systems can operate by themselves as off-grid PV systems, it focuses on systems connected to the utility grid, or grid-tied PV systems.

Types of PV Systems

Photovoltaic-based systems are generally classified according to their functional and operational requirements, their component configuration, and how the equipment is connected to the other power sources and electrical loads (appliances). The two principle classifications are grid-connected and stand alone systems.

Grid Connected

Grid-connected or utility-intertie PV systems are designed to operate in parallel with and interconnected with the electric utility grid. The primary component is the inverter, or power conditioning unit (PCU). The inverter converts the DC power produced by the PV array into AC power consistent with the voltage and power quality required by the utility grid. The inverter automatically stops supplying power to the grid when the utility grid is not energized. A bi-directional interface is made between the PV system AC output circuits and the electric utility network, typically at an on-site distribution panel or service entrance.

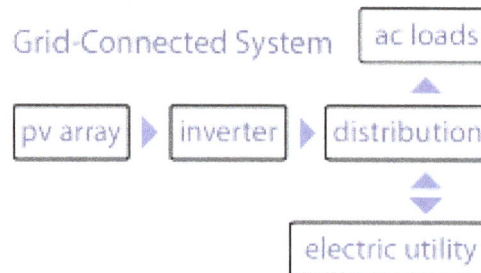

Grid-Connected System   [ac loads]

[pv array] ▶ [inverter] ▶ [distribution]

[electric utility]

This allows the power produced by the PV system to either supply on-site electrical loads, or to back feed the grid when the PV system output is greater than the on-site load demand. During periods when the electrical demand is greater than the PV system output (night-time), the balance of power required is received from the electric utility This safety feature is required in all grid-connected PV systems, it also ensures that the PV system will not continue to operate and feed back onto the utility grid when the grid is down for service or repair.

Stand Alone System

Stand-alone PV systems are designed to operate independent of the electric utility grid, and are generally designed and sized to supply certain DC and/or AC electrical loads. Stand-alone systems may be powered by a PV array only, or may use wind, an engine-generator or utility power as a backup power source in what is called a PV-hybrid system. The simplest type of stand-alone PV system is a direct-coupled system, where the DC output of a PV module or array is directly connected to a DC load.

Direct-coupled Stand Alone System

[pv array] ▶ ▶ ▶ ▶ [dc loads]

Since there is no electrical energy storage (batteries) in direct-coupled systems, the load only operates during sunlight hours, making these designs suitable for common applications such as ventilation fans, water pumps, and small circulation pumps for solar thermal water heating systems. Matching the impedance of the electrical load to the maximum power output of the PV array is a critical part of designing well-performing direct-coupled system. For certain loads such as positive-displacement water pumps; a type of electronic DC-DC converter, called a maximum power point tracker (MPPT) is used between the array and load to help better utilize the available array maximum power output.

In many stand-alone PV systems, batteries are used for energy storage. Below is a diagram of a typical stand-alone PV system with battery storage powering DC and AC loads.

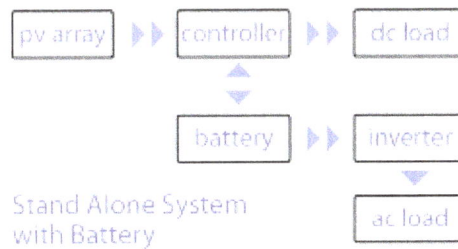

Stand Alone System
with Battery

Below is a diagram of a photovoltaic hybrid system with battery storage powering DC and AC loads and using a using a backup power source (wind, engine-generator or utility power).

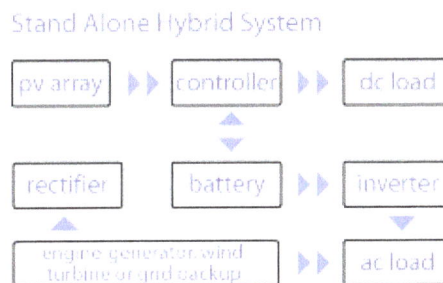

Stand Alone Hybrid System

Steps in Designing a PV System

Calculate the Electrical Load

Examine the uses of energy in a home in three categories (thermal or heat energy, electrical energy, and refrigeration), conservation opportunities can then be isolated in each category that can affect overall electrical consumption.

Thermal Energy Requirement for Heating Living Spaces, Water, and Cooking

Best accomplished by non-electrical fuels such as solar, gas, wood, and others. Electric space heating, water heating, and cooking require an enormous amount of electricity. It is not practical to use photovoltaics to create electricity for these purposes. Solar energy can be used in other forms such as passive and active solar space heating and solar water heating more efficiently. Gas can also be used for the thermal loads more economically and efficiently than electricity.

Electrical Loads (Lighting, Appliance and Equipment Operation)

Should be done with the most conserving items that can accomplish the task. Highly energy efficient lighting products are readily available and the energy efficiency of appliances can be easily compared for the best choices. Best application for PVs is in this category.

Refrigeration for Air Conditioning and Food Preservation

Consumes proportionally enormous amounts of electrical energy making PV power very costly for these tasks. Gas powered air conditioning is available as an alternative.

For food preservation, there are gas refrigerators and two manufacturers of very high efficiency electrical refrigerators and freezers.

Size the PV System

Different size PV panels will produce different amounts of power. The rated output wattage of the panel is the amount of watts the panel will create in one hour of direct sun. For our area, multiply the rated wattage by 5.1 to get the average amount produced in one day. The 5.1 factor is the viable operating hours per day and accounts for the fact that there will be more sun available in the summer and less in the winter.

Example

If a panel is rated at 48 watts, multiply that figure by 5.1 to get 245 watt-hours per day. Use that figures divided into the "Daily Energy Use" that was calculated above and the resulting number will be the number of panels of that particular size you will need. If the "Daily Energy Use" figure above was 2,000 watts per day, 2,000 divided by 245 gives us 8.16, rounded up to 9 panels. (Note that there are tracking systems that will increase the effective hours of sunlight striking a PV panel beyond 5.1.)Panel Rating (48) x Avg. operating time (5.1) = panel watt-hours per day (245).

Daily Energy Use (2,000) / Panel watt-hours (245)= number of panels (8.16), round up to even number = 9.

Photovoltaic Systems Work

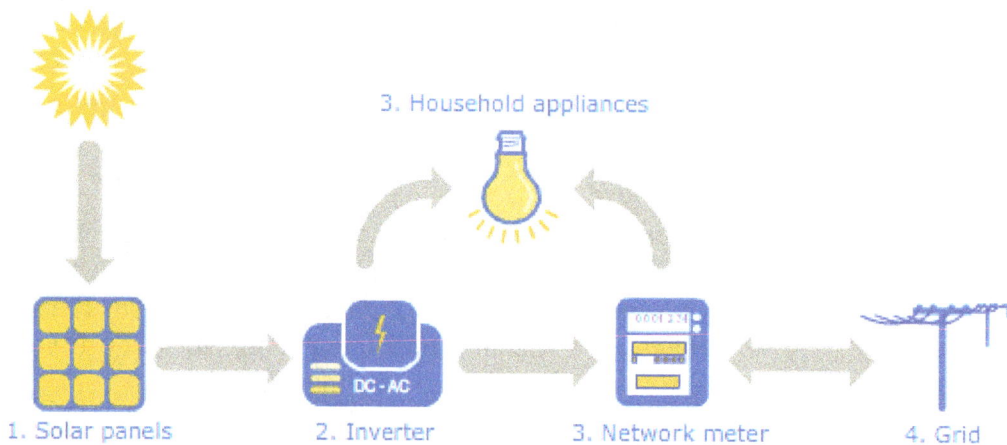

Figure: Residential grid-tied solar PV system diagram

The light from the Sun, made up of packets of energy called photons, falls onto a solar panel and creates an electric current through a process called the photovoltaic effect. Each panel produces a relatively small amount of energy, but can be linked together with other panels to produce higher amounts of energy as a solar array. The electricity produced from a solar panel (or array) is in the form of direct current (DC). Although many electronic devices use DC electricity, including your phone or laptop, they are designed to operate using the electrical utility grid which provides (and requires) alternating current (AC). Therefore, in order for the solar electricity to be useful it must first be converted from DC to AC using an inverter. This AC electricity from the inverter can then be used to power electronics locally, or be sent on to the electrical grid for use elsewhere.

System Components

In addition to the solar panels, there are other important components of a photovoltaic system which are commonly referred to as the "balance of system" or BOS. These components (which typically account for over half of the system cost and most the of maintenance) can include inverters, racking, wiring, combiners, disconnects, circuit breakers and electric meters.

Solar Panel

Figure: A solar panel, consisting of many photovoltaic cells.

A solar panel consists of many solar cells with semiconductor properties encapsulated within a material to protect it from the environment. These properties enable the cell to capture light, or more specifically, the photons from the sun and convert their energy into useful electricity through a process called the photovoltaic effect. On either side of the semiconductor is a layer of conducting material which "collects" the electricity produced. The illuminated side of the panel also contains an anti-reflection coating to minimize the losses due to reflection. The majority of solar panels produced worldwide are made from crystalline silicon, which has a theoretical efficiency limit of 33% for converting the Sun's energy into electricity. Many other semiconductor materials and solar cell technologies have been developed that operate at higher efficiencies, but these come with a higher cost to manufacture.

Inverters

An inverter is an electrical device which accepts electrical current in the form of direct current (DC) and converts it to alternating current (AC). For solar energy systems, this means the DC current from the solar array is fed through an inverter which converts it to AC. This conversion is necessary to operate most electric devices or interface with the electrical grid. Inverters are important for almost all solar energy systems and are typically the most expensive component after the solar panels themselves.

Figure: A solar inverter (yellow) mounted to the solar racking converts DC electricity from the solar array to useful AC electricity.

Most inverters have conversion efficiencies of 90% or higher and contain important safety features including ground fault circuit interruption and anti-islanding. These shut down the PV system when there is a loss of grid power.

Racking

Racking refers to the mounting apparatus which fixes the solar array to the ground or rooftop. Typically constructed from steel or aluminum, these apparatuses mechanically fix the solar panels in place with a high level of precision. Racking systems should be designed to withstand extreme weather events such as hurricane or tornado level wind speeds and/or high accumulations of snow. Another important feature of racking systems is to electrically bond and ground the solar array to prevent electrocution. Rooftop racking systems typically come in two variations including flat roof systems and pitched roof systems. For flat rooftops it is common for the racking system to include weighted ballast to hold the array to the roof using gravity. On pitched rooftops, the racking system must be mechanically anchored to the roof structure. Ground mounted PV systems, as shown in figure, can also use either ballast or mechanical anchors to fix the array to the ground. Some ground mounted racking systems also incorporate tracking systems which use motors and sensors to track the Sun through the sky, increasing the amount of energy generated at a higher equipment and maintenance cost.

Other Components

The remaining components of a typical solar PV system include combiners, disconnects, breakers, meters and wiring. A solar combiner, as the name suggests, combines two or more electrical cables into one larger one. Combiners typically include fuses for protection and are used on all medium to large and utility-scale solar arrays. Disconnects are electrical gates or switches which allow for manual disconnection of an electrical wire. Typically used on either side of an inverter, namely the "DC disconnect" and "AC disconnect" these devices provide electrical isolation when an inverter needs to be installed or replaced. Circuit breakers or breakers protect electrical systems from over current or surges. Designed to trigger automatically when the current reaches a predetermined amount, breakers can also be operated manually, acting as an additional disconnect. An electric meter measures the amount of energy that passes through it and is commonly used by electric utility companies to measure and charge customers. For solar PV systems, a special bi-directional electric meter is used to measure both the incoming energy from the utility, and the outgoing energy from the solar PV system. Finally, the wiring or electrical cables transport the electrical energy from and between each component and must be properly sized to carry the current. Wiring exposed to sunlight must have protection against UV exposure, and wires carrying DC current sometimes require metal sheathing for added protection.

Advantages of a Photovoltaic System

Photovoltaic systems are very efficient, and installing them comes with the following benefits:

1. The solar energy they produce is sustainable. This is because the sun is a renewable resource and cannot be over-consumed. Hence, it is an abundant and natural source.

2. Solar energy can be produced anywhere. Thus, even though countries closer to the Equator have greater potential for producing it, it is not available only to them. It can be produced anywhere, even on cloudy days. Hence its availability makes it very valuable and useful.

3. Solar energy is free. Even though the installation costs are high, you will not be charged for the electricity that the photovoltaic system will provide you later on. Hence, solar power is a free electricity resource.

4. Using solar energy will reduce your electricity bills in the long run. In some cases, you can also generate income by selling your surplus energy back to your energy provider.

5. Having a photovoltaic system installed in your house will increase your independence and the value of your home. Thus, you will not be dependent on nonrenewable energy resources that you used before.

6. The more homes that use solar energy, the more jobs will be created in the renewable energy resources. Hence, you will contribute to a wider job market.

Despite producing electricity, solar energy can be used to heat water or to provide heat for homes and businesses. Thus, solar thermal and solar water heating can be interesting to look at, if you want to include solar energy all around your house. What is more, using the newest technologies, such as thermodynamic panels, water can be heated by simply exploiting the temperature of the ambient air.

Disadvantages of a Photovoltaic System

*   A large area of unshaded south, south-west or south-east facing roof is required to maximize payback. Smaller systems can be installed but payback will be longer.

*   Panels degrade over time by approximately 20% over 25 years; this however is taken into account in most reputable suppliers calculations.

*   It may be beneficial to replace the inverter after 10 years to optimize power generation, although this is not essential.

Application

Lighting

With the invention of LED (light emitting diode) technology as low power lighting sources, PV systems find an ideal application in remote or mobile lighting systems. PV systems combined with battery storage facilties are mostly used to provide lighting for billboards, highway in formation signs, public-use facilities, parking lots, vacation cabins, lighting for trains.

Figure: A street light powered by PV

Figure: Portable lighting system along with mobile charging facility

Figure: Example of a bill board powered by solar panels

Communications

Signals required by communication systems need amplification after particular distance intervals. Various relay towers are stationed to boost radio, television, and phone signals. High grounds are mostly favoured as the sites for repeater stations. These sites are generally far from power lines. To reduce the difficulty and cost associated with generators, PV systems are being installed as a viable alternative.

Figure: Repeater at an elevated location can be powered by PV.

Figure: Satellites used for communication are powered by PV.

Electricity for Remote Areas

Some areas are quite far from the distribution network to establish connection with the grid. Areas under construction also need power supply before they are connected. PV systems are an attractive option for these cases. Furthermore, PV systems can be backed up by conventional generators to provide uninterrupted supply.

Figure: Remote area in Africa powered by PV.

Disaster Relief

Natural calamities often bring about an electricity crisis. As the disasters such as hurricanes, floods, tornadoes, and earthquakes destroy electricity generation and distribution systems. In situations like these, where power will be out for an extended period, portable PV system can provide temporary solutions for light, communication, food and water systems. Emergency health clinics opt for PV based electricity over conventional systems in lieu to problems of fuel transport and pollution.

Figure: Portable PV systems powering area struck by natural disaster.

Scientific Experiments

In various cases, scientific experiments are set up in areas far from power supply. PV systems can be effectively used to carry out scientific activities in remote areas. Systems monitoring seismic activities, highway conditions, meteorological information and other research activities can be powered by PV systems.

Figure: Remote scientific experiment powered by PV.

Signal Systems

Navigational systems, such as light houses, highway and aircraft warning signals can be far from the electric grid. PV systems can be a reliable power source for these important applications. Even portable traffic lights can be powered by PV systems.

Figure: Light house being powered by PV.

Figure: Portable traffic lights powered by PV.

Water Pumping

PV is a perfect candidate for agricultural and livestock purposes due to the need for water during the periods with bright sunshine. These pumping systems can supply water directly to fields, or can store water for the time of need. These systems can even be used to provide water to remote areas and villages.

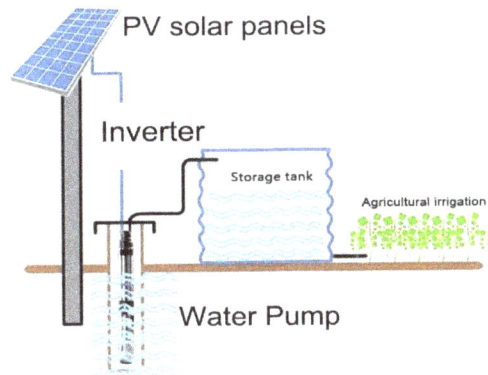

Figure: Solar water pumps can be very cost effective for remote agricultural activities.

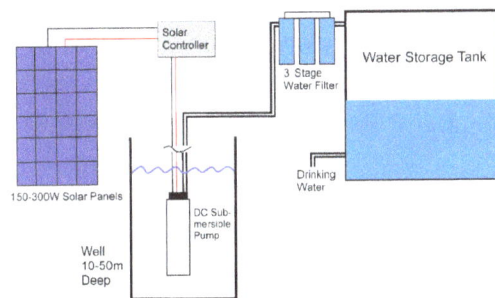

Figure: Solar water pumps can supply water to areas with no connection to grid.

Charging Vehicle Batteries

Vehicles running on electric power can be charged at PV powered stations. Such vehicles can also maintain their critical battery states using PV powered sources. Boats and other leisure vehicles can be charged directly using PV systems.

Figure: Electric vehicles at PV charging station.

Solar Power Cathodic Protection

Pipelines, well heads and other metallic structures are prone to corrosion due to exposure to water. Corrosion occurs due to the electrolytic activity of metals as they lose ions in contact with water. This electrolytic process leading to corrosion can be reduced by applying an external voltage. This external voltage will prevent the ion loss from the metal. To that end, only a small DC voltage will be enough. PV is a suitable candidate for this purpose as they produce low voltage DC power that can be used directly.

Figure: Cathodic protection using PV.

Refrigeration

PV system can be exceptionally suitable for storage and transport of medicines and vaccines that require refrigeration.

Figure: Solar powered refrigerator.

Consumer Products

PV technology is being used for variety of commercially available consumer based products. Small DC appliances such as toys, watches, calculators, radios, televisions, flashlights, fans etc. can operate with PV based energy systems.

Figure: Solar powered calculator and radio.

Public Utilities

Various public utility systems such as teller machines and telephone booths can also be powered by PV systems.

Figure: ATM and telephone booth powered by PV.

Photovoltaic Cell

A solar cell made from a monocrystalline silicon wafer.

A solar cell or photovoltaic cell is a device that converts solar energy into electricity by the photovoltaic effect. Sometimes, the term *solar cell* is reserved for devices intended specifically to capture energy from sunlight, while the term *photovoltaic cell* is used when the source is unspecified. Assemblies of cells are used to make solar panel, solar modules, or photovoltaic arrays. Photovoltaics is the field of technology and research related to the application of solar cells for solar energy.

Three Generations of Solar Cells

Solar cells are classified into three generations which indicates the order of which each became important. At present there is concurrent research into all three generations while the first generation technologies are most highly represented in commercial production, accounting for 89.6 percent of 2007 production.

First Generation

First generation cells consist of large-area, high quality and single junction devices. First generation technologies involve high energy and labor inputs which prevent any significant progress in reducing production costs. Single junction silicon devices are approaching the theoretical limiting efficiency of 33 percent and achieve cost parity with fossil fuel energy generation after a payback period of 5-7 years. They are not likely to get lower than US$1/W.

Second Generation

Second generation materials have been developed to address energy requirements and production costs of solar cells. Alternative manufacturing techniques such as vapor deposition, electroplating, and use of Ultrasonic Nozzles are advantageous as they reduce high temperature processing significantly. It is commonly accepted that as manufacturing techniques evolve production costs will be dominated by constituent material requirements, whether this be a silicon substrate, or glass cover. Second generation technologies are expected to gain market share in 2008.

Such processes can bring costs down to a little under US$0.50/W but because of the defects inherent in the lower quality processing methods, have much reduced efficiencies compared to First Generation.

The most successful second generation materials have been cadmium telluride (CdTe), copper indium gallium selenide, amorphous silicon and micromorphous silicon. These materials are applied in a thin film to a supporting substrate such as glass or ceramics reducing material mass and therefore costs. These technologies do hold promise of higher conversion efficiencies, particularly CIGS-CIS, DSC and CdTe offers significantly cheaper production costs.

Among major manufacturers, there is certainly a trend toward second generation technologies however commercialization of these technologies has proven difficult. In 2007, First Solar produced 200 MW of CdTe solar cells making it the fifth largest producer of solar cells in 2007 and the first ever to reach the top ten from production of second generation technologies alone. Honda Soltec Co., Ltd., a Honda company, also began to commercialize] their CIGS base solar panel in 2008.

In 2007 CdTe production represented 4.7 percent of total market share, thin-film silicon 5.2 percent and CIGS 0.5 percent.

Third Generation

Third generation technologies aim to enhance poor electrical performance of second generation (thin-film technologies) while maintaining very low production costs.

Current research is targeting conversion efficiencies of 30-60 percent while retaining low cost materials and manufacturing techniques. They can exceed the theoretical solar conversion efficiency limit for a single energy threshold material that was calculated in 1961 by Shockley and Queisser as 31 percent under 1 sun illumination and 40.8 percent under maximal concentration of sunlight (46,200 suns, which makes the latter limit more difficult to approach than the former).

There are a few approaches to achieving these high efficiencies:

- Multi junction photovoltaic cell (multiple energy threshold devices).

- Modifying incident spectrum (concentration).
- Use of excess thermal generation (caused by UV light) to enhance voltages or carrier collection.
- Use of infrared spectrum to produce electricity at night.

Technologies include:

- Silicon nanostructures
- Up/Down converters
- Hot-carrier cells
- Thermoelectric cells

High Efficiency Cells

High efficiency solar cells are a class of solar cells that can generate electricity at higher efficiencies than conventional solar cells. While high efficiency solar cells are more efficient in terms of electrical output per incident energy (watt/watt), much of the industry is focused on the most cost efficient technologies (cost-per-watt or $/watt). Still, many businesses and academics are focused on increasing the electrical efficiency of cells, and much development is focused on high efficiency solar cells.

Records

Monocrystalline Si

In 1994, the University of New South Wales (UNSW) reported the highest silicon solar cell efficiency of 24.7 percent with their PERL cell technology. This record was valid until 2008.

UNSW's ARC Photovoltaic Centre of Excellence reported the first silicon solar cell to achieve the milestone of 25 per cent efficiency.

Polycrystalline Si

Multiple Junction Solar Cells

The record for multiple junction solar cells is disputed. A team lead by the University of Delaware, the Fraunhofer Institute, and NREL all claim the world record title at 42.8, 41.1, and 40.8 percent, respectively. NREL claims that the other implementations have never been put under any standardized tests and, in the case of the University of Delaware project, represents only hypothetical efficiencies of a panel that's never been fully assembled.

Thin Film Solar Cells

In 2002, the highest reported efficiency for solar cells based on thin films of CdTe is 18 percent, which was achieved by the research group of prof. I.M.Dharmadasa at Sheffield Hallam University in the United Kingdom.

The US national renewable energy research facility NREL achieved an efficiency of 19.9 percent for the solar cells based on copper indium gallium selenide thin films, also known as CIGS. These CIGS films have been grown by physical vapor deposition in a three-stage co-evaporation process. In this process In, Ga and Se are evaporated in the first step; in the second step it is followed by Cu and Se co-evaporation and in the last step terminated by In, Ga and Se evaporation again.

Applications and Implementations

Polycrystaline PV cells laminated to backing material in a PV module

Polycrystalline PV cells

Solar cells are often electrically connected and encapsulated as a module. PV modules often have a sheet of glass on the front (sun up) side, allowing light to pass while protecting the semiconductor wafers from the elements (rain, hail, etc.). Solar cells are also usually connected in series in modules, creating an additive voltage. Connecting cells in parallel will yield a higher current. Modules are then interconnected, in series or parallel, or both, to create an array with the desired peak DC voltage and current.

The power output of a solar array is measured in watts or kilowatts. In order to calculate the typical energy needs of the application, a measurement in watt-hours, kilowatt-hours or kilowatt-hours per day is often used. A common rule of thumb is that average power is equal to 20 percent of peak power, so that each peak kilowatt of solar array output power corresponds to energy production of 4.8 kWh per day (24 hours × 1kWh × 20 percent = 4.8 kWh)

To make practical use of the solar-generated energy, the electricity is most often fed into the electricity grid using inverters (grid-connected PV systems); in stand alone systems, batteries are used to store the energy that is not needed immediately.

Theory

1. Photons in sunlight hit the solar panel and are absorbed by semiconducting materials, such as silicon.

2. Electrons (negatively charged) are knocked loose from their atoms, allowing them to flow through the material to produce electricity. Due to the special composition of solar cells, the electrons are only allowed to move in a single direction. The complementary positive charges that are also created (like bubbles) are called holes and flow in the direction opposite of the electrons in a silicon solar panel.

3. An array of solar cells converts solar energy into a usable amount of direct current (DC) electricity.

Photogeneration of Charge Carriers

When a photon hits a piece of silicon, one of three things can happen:

1. The photon can pass straight through the silicon—this (generally) happens for lower energy photons,

2. The photon can reflect off the surface,

3. The photon can be absorbed by the silicon, if the photon energy is higher than the silicon band gap value. This generates an electron-hole pair and sometimes heat, depending on the band structure.

When a photon is absorbed, its energy is given to an electron in the crystal lattice. Usually this electron is in the valence band, and is tightly bound in covalent bonds between neighboring atoms, and hence unable to move far. The energy given to it by the photon "excites" it into the conduction band, where it is free to move around within the semiconductor. The covalent bond that the electron was previously a part of now has one fewer electron—this is known as a hole. The presence of a missing covalent bond allows the bonded electrons of neighboring atoms to move into the "hole," leaving another hole behind, and in this way a hole can move through the lattice. Thus, it can be said that photons absorbed in the semiconductor create mobile electron-hole pairs.

A photon need only have greater energy than that of the band gap in order to excite an electron from the valence band into the conduction band. However, the solar frequency spectrum approximates a black body spectrum at ~6000K, and as such, much of the solar radiation reaching the Earth is composed of photons with energies greater than the band gap of silicon. These higher energy photons will be absorbed by the solar cell, but the difference in energy between these photons and the silicon band gap is converted into heat (via lattice vibrations—called phonons) rather than into usable electrical energy.

Charge Carrier Separation

There are two main modes for charge carrier separation in a solar cell:

1. Drift of carriers, driven by an electrostatic field established across the device

2. Diffusion of carriers from zones of high carrier concentration to zones of low carrier concentration (following a gradient of electrochemical potential).

In the widely used p-n junction solar cells, the dominant mode of charge carrier separation is by drift. However, in non-p-n-junction solar cells (typical of the third generation solar cell research such as dye and polymer solar cells), a general electrostatic field has been confirmed to be absent, and the dominant mode of separation is via charge carrier diffusion.

p-n Junction

The most commonly known solar cell is configured as a large-area p-n junction made from silicon. As a simplification, one can imagine bringing a layer of n-type silicon into direct contact with a layer of p-type silicon. In practice, p-n junctions of silicon solar cells are not made in this way, but rather, by diffusing an n-type dopant into one side of a p-type wafer (or vice versa).

If a piece of p-type silicon is placed in intimate contact with a piece of n-type silicon, then a diffusion of electrons occurs from the region of high electron concentration (the n-type side of the junction) into the region of low electron concentration (p-type side of the junction). When the electrons diffuse across the p-n junction, they recombine with holes on the p-type side. The diffusion of carriers does not happen indefinitely however, because of an electric field which is created by the imbalance of charge immediately on either side of the junction which this diffusion creates. The electric field established across the p-n junction creates a diode that promotes current in only one direction across the junction. Electrons may pass from the n-type side into the p-type side, and holes may pass from the p-type side to the n-type side, but not the other way around. This region where electrons have diffused across the junction is called the depletion region because it no longer contains any mobile charge carriers. It is also known as the "space charge region."

Connection to an External Load

Ohmic metal-semiconductor contacts are made to both the n-type and p-type sides of the solar cell, and the electrodes connected to an external load. Electrons that are created on the n-type side, or have been "collected" by the junction and swept onto the n-type side, may travel through the wire, power the load, and continue through the wire until they reach the p-type semiconductor-metal contact. Here, they recombine with a hole that was either created as an electron-hole pair on the p-type side of the solar cell, or are swept across the junction from the n-type side after being created there.

The voltage measured is equal to the difference in the quasi Fermi levels of the minority carriers, i.e., electrons in the p-type portion, and holes in the n-type portion.

Equivalent Circuit of a Solar Cell

To understand the electronic behavior of a solar cell, it is useful to create a model which is electrically equivalent, and is based on discrete electrical components whose behavior is well known. An ideal solar cell may be modeled by a current source in parallel with a diode; in practice no solar cell is ideal, so a shunt resistance and a series resistance component are added to the model. The resulting equivalent circuit of a solar cell is shown below. Also shown, is the schematic representation of a solar cell for use in circuit diagrams.

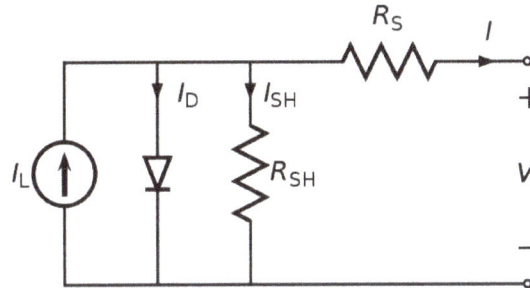

The equivalent circuit of a solar cell

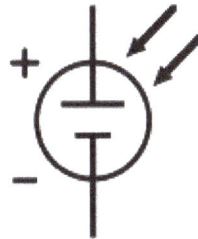

Schematic symbol of a solar cell

Characteristic Equation

From the equivalent circuit, it is evident that the current produced by the solar cell is equal to that produced by the current source, minus that which flows through the diode, minus that which flows through the shunt resistor:

$$I = I_L - I_D - I_{SH}$$

Where

- I = output current (amperes),
- I_L = photogenerated current (amperes),
- I_D = diode current (amperes),
- I_{SH} = shunt current (amperes).

The current flowing through these elements governed by the voltage across them:

$$V_j = V + IR_S$$

where

- V = voltage across the output terminals (volts),
- I = output current (amperes),
- R_S = series resistance (Ω).

By the Shockley diode equation, the current diverted through the diode is:

$$I_D = I_0 \left\{ \exp\left[\frac{qV_j}{nkT}\right] - 1 \right\}$$

Where

- I_0 = reverse saturation current (amperes),
- n = diode ideality factor (1 for an ideal diode),
- q = elementary charge,
- k = Boltzmann's constant,
- T = absolute temperature,

For silicon at 25°C, $kT/q \approx 0.0259$ volts.

By Ohm's law, the current diverted through the shunt resistor is:

$$I_{SH} = \frac{V_j}{R_{SH}}$$

Where

- R_{SH} = shunt resistance (Ω).

Substituting these into the $I = I_L - I_D - I_{SH}$ produces the characteristic equation of a solar cell, which relates solar cell parameters to the output current and voltage:

$$I = I_L - I_0 \left\{ \exp\left[\frac{q(V + IR_S)}{nkT}\right] - 1 \right\} - \frac{V + IR_S}{R_{SH}}$$

An alternative derivation produces an equation similar in appearance, but with V on the left-hand side. The two alternatives are identities; that is, they yield precisely the same results.

In principle, given a particular operating voltage V the equation may be solved to determine the operating current I at that voltage. However, because the equation involves I on both sides in a transcendental function the equation has no general analytical solution. However, even without a solution it is physically instructive. Furthermore, it is easily solved using numerical methods. (A general analytical solution to the equation is possible using Lambert's W function, but since Lambert's W generally itself must be solved numerically this is a technicality.)

Since the parameters I_0, n, R_S, and R_{SH} cannot be measured directly, the most common application of the characteristic equation is nonlinear regression to extract the values of these parameters on the basis of their combined effect on solar cell behavior.

Effect of Physical Size

The values of I_0, R_S, and R_{SH} are dependent upon the physical size of the solar cell. In comparing otherwise identical cells, a cell with twice the surface area of another will, in principle, have double the I_0 because it has twice the junction area across which current can leak. It will also have half the R_S and R_{SH} because it has twice the cross-sectional area through which current can flow. For this reason, the characteristic equation is frequently written in terms of current density, or current produced per unit cell area:

$$J = J_L - J_0 \left\{ \exp \left[\frac{q(V + Jr_S)}{nkT} \right] - 1 \right\} - \frac{V + Jr_S}{r_{SH}}$$

Where

- J = current density (amperes/cm²)
- J_L = reverse saturation current density (amperes/cm²)
- r_S = specific series resistance (Ω-cm²)
- r_{SH} = specific shunt resistance (Ω-cm²)

This formulation has several advantages. One is that since cell characteristics are referenced to a common cross-sectional area they may be compared for cells of different physical dimensions. While this is of limited benefit in a manufacturing setting, where all cells tend to be the same size, it is useful in research and in comparing cells between manufacturers. Another advantage is that the density equation naturally scales the parameter values to similar orders of magnitude, which can make numerical extraction of them simpler and more accurate even with naive solution methods.

A practical limitation of this formulation is that as cell sizes shrink, certain parasitic effects grow in importance and can affect the extracted parameter values. For example, recombination and contamination of the junction tend to be greatest at the perimeter of the cell, so very small cells may exhibit higher values of J_0 or lower values of γ_{SH} than larger cells that are otherwise identical. In such cases, comparisons between cells must be made cautiously and with these effects in mind.

Cell Temperature

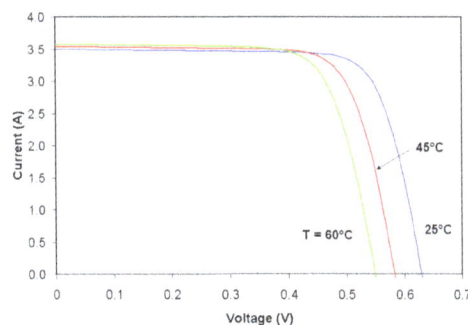

Effect of temperature on the current-voltage characteristics of a solar cell

Temperature affects the characteristic equation in two ways: directly, via T in the exponential term, and indirectly via its effect on I_0. (Strictly speaking, temperature affects all of the terms, but these two far more significantly than the others.) While increasing T reduces the magnitude of the exponent in the characteristic equation, the value of I_0 increases in proportion to $\exp T$. The net effect is to reduce V_{OC} linearly with increasing temperature. The magnitude of this reduction is inversely proportional to V_{OC}; that is, cells with higher values of V_{OC} suffer smaller reductions in voltage with increasing temperature. For most crystalline silicon solar cells the reduction is about 0.50%/°C, though the rate for the highest-efficiency crystalline silicon cells is around 0.35%/°C. By way of comparison, the rate for amorphous silicon solar cells is 0.20-0.30%/°C, depending on how the cell is made.

The amount of photogenerated current I_L increases slightly with increasing temperature because of an increase in the number of thermally generated carriers in the cell. This effect is slight, however: about 0.065%/°C for crystalline silicon cells and 0.09 percent for amorphous silicon cells.

The overall effect of temperature on cell efficiency can be computed using these factors in combination with the characteristic equation. However, since the change in voltage is much stronger than the change in current, the overall effect on efficiency tends to be similar to that on voltage. Most crystalline silicon solar cells decline in efficiency by 0.50%/°C and most amorphous cells decline by 0.15-0.25%/°C. The figure above shows I-V curves that might typically be seen for a crystalline silicon solar cell at various temperatures.

Series Resistance

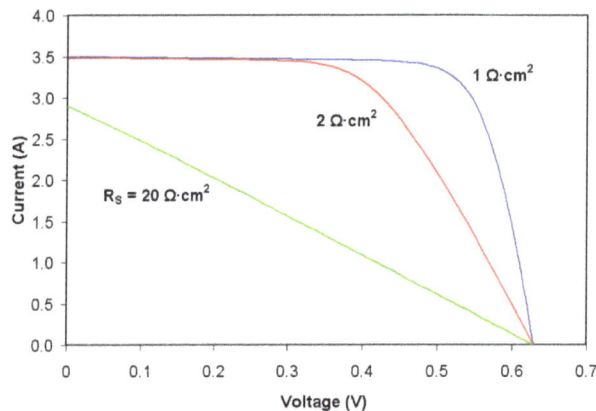

Effect of series resistance on the current-voltage characteristics of a solar cell

As series resistance increases, the voltage drop between the junction voltage and the terminal voltage becomes greater for the same flow of current. The result is that the current-controlled portion of the I-V curve begins to sag toward the origin, producing a significant decrease in the terminal voltage V and a slight reduction in I_{sc}. Very high values of R_S will also produce a significant reduction in I_{sc}; in these regimes, series resistance dominates and the behavior of the solar cell resembles that of a resistor. These effects are shown for crystalline silicon solar cells in the I-V curves displayed in the figure above.

Shunt Resistance

As shunt resistance decreases, the flow of current diverted through the shunt resistor increases for a given level of junction voltage. The result is that the voltage-controlled portion of the I-V curve begins to sag toward the origin, producing a significant decrease in the terminal current I and a slight reduction in V_{oc}. Very low values of R_{SH} will produce a significant reduction in V_{oc}. Much as in the case of a high series resistance, a badly shunted solar cell will take on operating characteristics similar to those of a resistor. These effects are shown for crystalline silicon solar cells in the I-V curves displayed in the figure below.

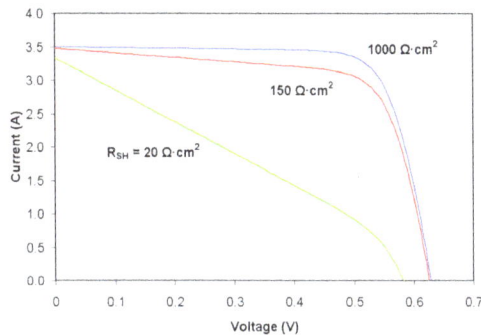

Effect of shunt resistance on the current-voltage characteristics of a solar cell

Reverse Saturation Current

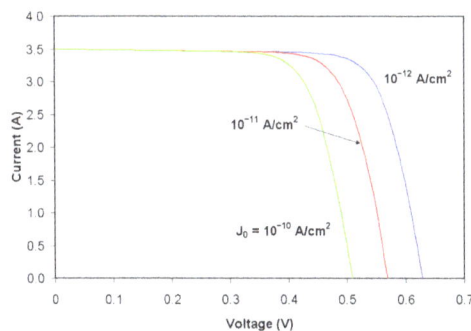

Effect of reverse saturation current on the current-voltage characteristics of a solar cell

If one assumes infinite shunt resistance, the characteristic equation can be solved for V_{oc}:

$$V_{oc} = \frac{kT}{q} \exp\left\{ \frac{I_{sc}}{I_0} + 1 \right\}$$

Thus, an increase in I_0 produces a reduction in V_{oc} proportional to the inverse of the logarithm of the increase. This explains mathematically the reason for the reduction in V_{oc} that accompanies increases in temperature. The effect of reverse saturation current on the I-V curve of a crystalline silicon solar cell are shown in the figure to the right. Physically, reverse saturation current is a measure of the "leakage" of carriers across the p-n junction in reverse bias. This leakage is a result of carrier recombination in the neutral regions on either side of the junction.

Ideality Factor

The ideality factor (also called the emissivity factor) is a fitting parameter that describes how closely the diode's behavior matches that predicted by theory, which assumes the p-n junction of the diode is an infinite plane and no recombination occurs within the space-charge region. A perfect match to theory is indicated when $n = 1$. When recombination in the space-charge region dominate other recombination, however, $n = 2$. The effect of changing ideality factor independently of all other parameters is shown for a crystalline silicon solar cell in the I-V curves displayed in the figure below.

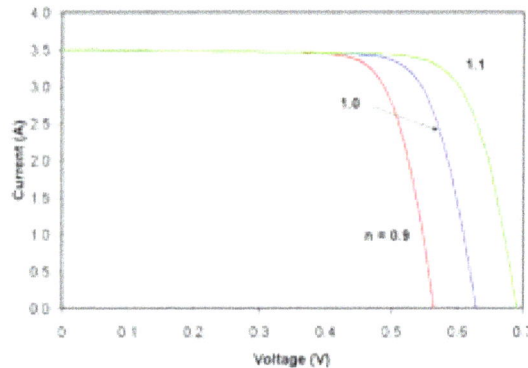

Effect of ideality factor on the current-voltage characteristics of a solar cell

Most solar cells, which are quite large compared to conventional diodes, well approximate an infinite plane and will usually exhibit near-ideal behavior under standard test condition ($n \approx 1$). Under certain operating conditions, however, device operation may be dominated by recombination in the space-charge region. This is characterized by a significant increase in I_0 as well as an increase in ideality factor to. The latter tends to erode solar cell output voltage while the former acts to increase it. The net effect, therefore, is a combination of the increase in voltage shown for increasing η and the decrease in voltage shown for increasing I_0 in the figure above. Typically, I_0 is the more significant factor and the result is a reduction in voltage.

Solar Cell Efficiency Factors

Energy Conversion Efficiency

A solar cell's *energy conversion efficiency* (η), is the percentage of power converted (from absorbed light to electrical energy) and collected, when a solar cell is connected to an electrical circuit. This term is calculated using the ratio of the maximum power point, P_m, divided by the input light *irradiance* (E, in W/m²) under standard test conditions (STC) and the *surface area* of the solar cell (A_c in m²).

$$\eta = \frac{P_m}{E \times A_c}$$

STC specifies a temperature of 25°C and an irradiance of 1000 W/m² with an air mass 1.5 (AM1.5) spectrum. These correspond to the irradiance and spectrum of sunlight incident on a clear day upon a sun-facing 37°-tilted surface with the sun at an angle of 41.81° above the horizon. This condition approximately represents solar noon near the spring and autumn equinoxes in the continental United States with surface of the cell aimed directly at the sun. Thus, under these conditions a solar cell of 12 percent efficiency with a 100 cm² (0.01 m²) surface area can be expected to produce approximately 1.2 watts of power.

The losses of a solar cell may be broken down into reflectance losses, thermodynamic efficiency, recombination losses and resistive electrical loss. The overall efficiency is the product of each of these individual losses.

Due to the difficulty in measuring these parameters directly, other parameters are measured instead: Thermodynamic efficiency, quantum efficiency, V_{OC} ratio, and fill factor. Reflectance losses are a portion of the quantum efficiency under "External quantum efficiency." Recombination losses make up a portion of the quantum efficiency, V_{OC} ratio, and Fill Factor. Resistive losses are predominantly categorized under Fill Factor, but also make up minor portions of the quantum efficiency, V_{OC} ratio.

Generally, solar cells on the market today do not produce much electricity from ultraviolet light, instead it is either filtered out or absorbed by the cell, heating the cell. That heat is wasted energy and could even lead to damage to the cell.

Thermodynamic Efficiency Limit

Solar cells operate as quantum energy conversion devices, and are therefore subject to the "Thermodynamic Efficiency Limit." Photons with an energy below the band gap of the absorber material cannot generate a hole-electron pair, and so their energy is not converted to useful output and only generates heat if absorbed. For photons with an energy above the band gap energy, only a fraction of the energy above the band gap can be converted to useful output. When a photon of greater energy is absorbed, the excess energy above the band gap is converted to kinetic energy of the carrier combination. The excess kinetic energy is converted to heat through phonon interactions as the kinetic energy of the carriers slows to equilibrium velocity.

Solar cells with multiple band gap absorber materials are able to more efficiently convert the solar spectrum. By using multiple band gaps, the solar spectrum may be broken down into smaller bins where the thermodynamic efficiency limit is higher for each bin.

Quantum Efficiency

As described above, when a photon is absorbed by a solar cell it can produce a pair of free charge carriers, i.e., an electron-hole pair. One of the carriers (the minority carrier) may then be able to reach the p-n junction and contribute to the current produced by the solar cell; such a carrier is said to be *collected*. Alternatively, the carrier may give up its energy and once again become bound to an atom within the solar cell *without being collected*; this process is then called *recombination* since one electron and one hole recombine and thereby annihilate the associated free charge. The carriers that recombine do *not contribute* to the generation of electrical current.

Quantum efficiency refers to the percentage of photons that are converted to electric current (i.e., collected carriers) when the cell is operated under short circuit conditions. *External* quantum efficiency (EQE) is the fraction of *incident* photons that are converted to electrical current, while *internal* quantum efficiency (IQE) is the fraction of *absorbed* photons that are converted to electrical current. Mathematically, internal quantum efficiency is related to external quantum efficiency by the reflectance (R) and the transmittance (T) of the solar cell by:

$$IQC = EQE/(1 - R - T).$$

Please note that for a thick bulk Si solar cell T is approximately zero and is therefore in practical cases often neglected.

Quantum efficiency should not be confused with energy conversion efficiency, as it does not convey information about the fraction of power that is converted by the solar cell. Furthermore, quantum efficiency is most usefully expressed as a *spectral* measurement (that is, as a function of photon wavelength or energy). Since some wavelengths are absorbed more effectively than others in most semiconductors, spectral measurements of quantum efficiency can yield valuable information about which parts of a particular solar cell design are most in need of improvement. Spectral quantum efficiency measurements also allow determining important solar cell parameters such as the base diffusion length.

Maximum-power Point

A solar cell may operate over a wide range of voltages (V) and currents (I). By increasing the resistive load on an irradiated cell continuously from zero (a *short circuit*) to a very high value (an *open circuit*) one can determine the maximum-power point, the point that maximizes V×I; that is, the load for which the cell can deliver maximum electrical power at that level of irradiation. (The output power is zero in both the short circuit and open circuit extremes).

A high quality, monocrystalline silicon solar cell, at 25°C cell temperature, may produce 0.60 volts open-circuit (Voc). The cell temperature in full sunlight, even with 25°C air temperature, will probably be close to 45°C, reducing the open-circuit voltage to 0.55 volts per cell. The voltage drops modestly, with this type of cell, until the short-circuit current is approached (Isc). Maximum power (with 45°C cell temperature) is typically produced with 75 percent to 80 percent of the open-circuit voltage (0.43 volts in this case) and 90 percent of the short-circuit current. This output can be up to 70 percent of the Voc x Isc product. The short-circuit current (Isc) from a cell is nearly proportional to the illumination, while the open-circuit voltage (Voc) may drop only 10 percent with a 80 percent drop in illumination. Lower-quality cells have a more rapid drop in voltage with increasing current and could produce only 1/2 Voc at 1/2 Isc. The usable power output could thus drop from 70 percent of the Voc x Isc product to 50 percent or even as little as 25 percent. Vendors who rate their solar cell "power" only as Voc x Isc, without giving load curves, can be seriously distorting their actual performance.

The maximum power point of a photovoltaic varies with incident illumination. For systems large enough to justify the extra expense, a maximum power point tracker tracks the instantaneous power by continually measuring the voltage and current (and hence, power transfer), and uses this information to dynamically adjust the load so the maximum power is *always* transferred, regardless of the variation in lighting.

Fill Factor

Another defining term in the overall behavior of a solar cell is the *fill factor (FF)*. This is the ratio of the *maximum power point* divided by the *open circuit voltage* (V_{oc}) and the *short circuit current* (I_{sc}):

$$FF = \frac{P_m}{V_{oc} \times I_{sc}} = \frac{\eta \times A_c \times E}{V_{oc} \times I_{sc}}$$

Comparison of Energy Conversion Efficiencies

At this point, discussion of the different ways to calculate efficiency for space cells and terrestrial cells is necessary to alleviate confusion. In space, where there is no atmosphere, the spectrum of the sun is relatively unfiltered. However, on earth, with air filtering the incoming light, the solar spectrum changes. To account for the spectral differences, a system was devised to calculate this filtering effect. Simply, the filtering effect ranges from Air Mass 0 (AM0) in space, to approximately Air Mass 1.5 on earth. Multiplying the spectral differences by the quantum efficiency of the solar cell in question will yield the efficiency of the device. For example, a Silicon solar cell in space might have an efficiency of 14 percent at AM0, but have an efficiency of 16 percent on earth at AM 1.5. Terrestrial efficiencies typically are greater than space efficiencies.

Solar cell efficiencies vary from 6 percent for amorphous silicon-based solar cells to 40.7 percent with multiple-junction research lab cells and 42.8 percent with multiple dies assembled into a hybrid package. Solar cell energy conversion efficiencies for commercially available *multi crystalline Si* solar cells are around 14-19 percent. The highest efficiency cells have not always been the most economical—for example a 30 percent efficient multi junction cell based on exotic materials such as gallium arsenide or indium selenide and produced in low volume might well cost one hundred times as much as an 8% efficient amorphous silicon cell in mass production, while only delivering about four times the electrical power.

However, there is a way to "boost" solar power. By increasing the light intensity, typically photogenerated carriers are increased, resulting in increased efficiency by up to 15 percent. These so-called "concentrator systems" have only begun to become cost-competitive as a result of the development of high efficiency GaAs cells. The increase in intensity is typically accomplished by using concentrating optics. A typical concentrator system may use a light intensity 6-400 times the sun, and increase the efficiency of a one sun GaAs cell from 31 percent at AM 1.5 to 35 percent.

A common method used to express economic costs of electricity-generating systems is to calculate a price per delivered kilowatt-hour (kWh). The solar cell efficiency in combination with the available irradiation has a major influence on the costs, but generally speaking the overall system efficiency is important. Using the commercially available solar cells (as of 2006) and system technology leads to system efficiencies between 5 and 19 percent. As of 2005, photovoltaic electricity generation costs ranged from ~0.60 US$/kWh (0.50 €/kWh) (central Europe) down to ~0.30 US$/kWh (0.25 €/kWh) in regions of high solar irradiation. This electricity is generally fed into the electrical grid on the customer's side of the meter. The cost can be compared to prevailing retail electric pricing (as of 2005), which varied from between 0.04 and 0.50 US$/kWh worldwide. In addition to solar irradiance profiles, these costs/kwh calculations will vary depending on assumptions for years of useful life of a system. Most c-Si panels are warranted for 25 years and should see 35+ years of useful life.

The chart below illustrates the various commercial large-area module energy conversion efficiencies and the best laboratory efficiencies obtained for various materials and technologies.

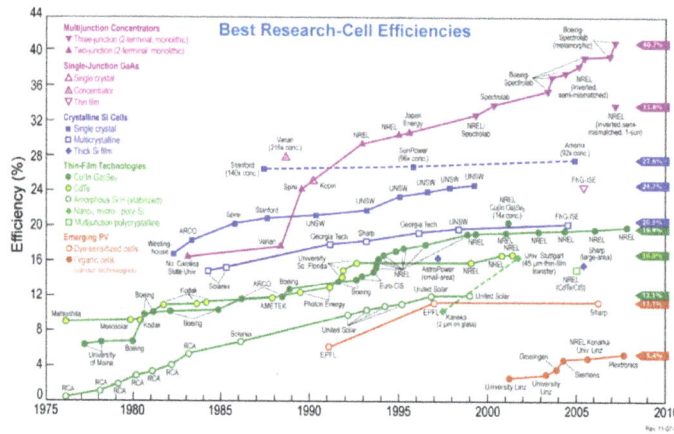

Reported timeline of solar cell energy conversion efficiencies (from National Renewable Energy Laboratory (USA)

Watts Peak

Since solar cell output power depends on multiple factors, such as the sun's incidence angle, for comparison purposes between different cells and panels, the measure of watts peak (Wp) is used. It is the output power under these conditions known as STC. Larger panels use the same rating system, but use kWp (1000 watts peak).

1. Insolation (solar irradiance) 1000 W/m²

2. Solar reference spectrum AM (airmass) 1.5

3. Cell temperature 25°C

Solar Cells and Energy payback

Energy payback is the recovery (period) of the energy spent for manufacturing of the respective technical energy systems, also called harvesting ratio (ISO 13602).

In the 1990s, when silicon cells were twice as thick, efficiencies were 30 percent lower than today and lifetimes were shorter, it may well have cost more energy to make a cell than it could generate in a lifetime. In the meantime, the technology has progressed significantly, and the energy payback time of a modern photovoltaic module is typically from 1 to 4 years depending on the type and where it is used. Generally, thin film technologies - despite having comparatively low conversion efficiencies - achieve significantly shorter energy payback times than conventional systems (often < 1 year). With a typical lifetime of 20 to 30 years, this means that modern solar cells are net energy producers, i.e., they generate significantly more energy over their lifetime than the energy expended in producing them.

Light-absorbing Materials

All solar cells require a *light absorbing material* contained within the cell structure to absorb photons and generate electrons via the *photoelectric effect*. The materials used in solar cells tend to have the property of preferentially absorbing the wavelengths of solar light that reach the earth surface; however, some solar cells are optimized for light absorption beyond Earth's atmosphere as well. Light absorbing materials can often be used in *multiple physical configurations* to take advantage of different light absorption and charge separation mechanisms.

Photovoltaic panels are normally made of either silicon or thin-film cells.

Many currently available solar cells are configured as bulk materials that are subsequently cut into wafers and treated in a "top-down" method of synthesis (silicon being the most prevalent bulk material).

Other materials are configured as thin-films (inorganic layers, organic dyes, and organic polymers) that are deposited on supporting substrates, while a third group are configured as nanocrystals and used as quantum dots (electron-confined nanoparticles) embedded in a supporting matrix in a "bottom-up" approach. Silicon remains the only material that is well-researched in both *bulk* (also called wafer-based) and *thin-film* configurations.

There are many new alternatives to silicon photocells. Proprietary nanoparticle silicon printing processes promises many of the photovoltaic features that conventional silicon can never achieve. It can be printed reel-to-reel on stainless steel or other high temperature substrates. However, most of the work on the next generation of photovoltaics is directed at printing onto low cost flexible polymer film and ultimately on common packaging materials. The main contenders are currently CIGS, CdTe, DSSC and organic photovoltaics.

Bulk

These *bulk* technologies are often referred to as wafer-based manufacturing. In other words, in each of these approaches, self-supporting wafers between 180 to 240 micrometers thick are processed and then soldered together to form a solar cell module.

Crystalline Silicon

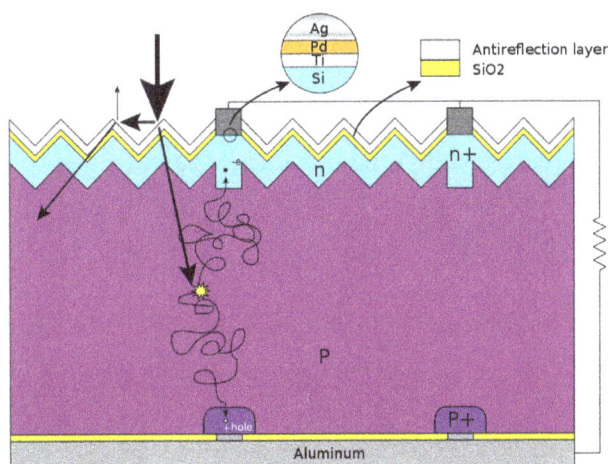

Basic structure of a silicon based solar cell and its working mechanism.

By far, the most prevalent *bulk* material for solar cells is crystalline silicon (*c-Si*), also known as "solar grade silicon." Bulk silicon is separated into multiple categories according to crystallinity and crystal size in the resulting ingot, ribbon, or wafer.

1. *Monocrystalline silicon* (c-Si): often made using the Czochralski process. Single-crystal wafer cells tend to be expensive, and because they are cut from cylindrical ingots, do not completely cover a square solar cell module without a substantial waste of refined silicon. Hence most *c-Si* panels have uncovered gaps at the four corners of the cells.

    Ribbon silicon is a type of monocrystalline silicon: it is formed by drawing flat thin films from molten silicon and having a multicrystalline structure. These cells have lower efficiencies than poly-Si, but save on production costs due to a great reduction in silicon waste, as this approach does not require sawing from ingots.

2. *Poly- or multicrystalline silicon* (poly-Si or mc-Si): made from cast square ingots—large blocks of molten silicon carefully cooled and solidified. Poly-Si cells are less expensive to produce than single crystal silicon cells, but are less efficient. US DOE data shows that there were a higher number of multicrystalline sales than monocrystalline silicon sales.

Thin Films

The various *thin-film* technologies currently being developed reduce the amount (or mass) of light absorbing material required in creating a *solar cell*. This can lead to reduced processing costs from that of bulk materials (in the case of silicon thin films) but also tends to reduce *energy conversion efficiency* (an average 7 to 10% efficiency), although many multi-layer thin films have efficiencies above those of bulk silicon wafers.

They have become popular compared to wafer silicon due to lower costs and advantages including flexibility, lighter weights, and ease of integration.

Cadmium Telluride Solar Cell

A cadmium telluride solar cell is a solar cell based on cadmium telluride, an efficient light-absorbing material for thin-film cells. Compared to other thin-film materials, CdTe is easier to deposit and more suitable for large-scale production.

Despite much discussion of the toxicity of CdTe-based solar cells, this is the only technology (apart from amorphous silicon) that can be delivered on a large scale.

Copper-indium Selenide

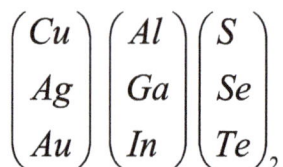

$$\begin{pmatrix} Cu \\ Ag \\ Au \end{pmatrix} \begin{pmatrix} Al \\ Ga \\ In \end{pmatrix} \begin{pmatrix} S \\ Se \\ Te \end{pmatrix}_2$$

Possible combinations of (I, III, VI) elements in the periodic table that have photovoltaic effect.

The materials based on $CuInSe_2$ that are of interest for photovoltaic applications include several elements from groups I, III and VI in the periodic table. These semiconductors are especially attractive for thin film solar cell application because of their high optical absorption coefficients and versatile optical and electrical characteristics which can in principle be manipulated and tuned for a specific need in a given device. CIS is an abbreviation for general chalcopyrite films of copper indium selenide ($CuInSe_2$), CIGS mentioned below is a variation of CIS. CIS films (no Ga) achieved greater than 14 percent efficiency. However, manufacturing costs of CIS solar cells at present are high when compared with amorphous silicon solar cells but continuing work is leading to more cost-effective production processes. The first large-scale production of CIS modules was started in 2006 in Germany by Wuerth Solar.

When gallium is substituted for some of the indium in CIS, the material is referred to as CIGS, or copper indium/gallium diselenide, a solid mixture of the semiconductors $CuInSe_2$ and $CuGaSe_2$, often abbreviated by the chemical formula $CuIn_xGa_{(1-x)}Se_2$. Unlike the conventional silicon based solar cell, which can be modeled as a simple p-n junction, these cells are best described by a more complex hetero junction model. Higher efficiencies (around 30 percent) can be obtained by using optics to concentrate the incident light or by using multi-junction tandem solar cells. The use of gallium increases the optical band gap of the CIGS layer as compared to pure CIS, thus increasing the open-circuit voltage, but decreasing the short circuit current.

In another point of view, gallium is added to replace indium due to gallium's relative availability to indium. Approximately 70 percent of indium currently produced is used by the flat-screen monitor industry. However, the atomic ratio for Ga in the >19 percent efficient CIGS solar cells is ~7 percent, which corresponds to a band gap of ~1.15 eV. CIGS solar cells with higher Ga amounts have lower efficiency. For example, CGS solar cells (which have a band gap of ~1.7eV have a record efficiency of 9.5 percent for pure CGS and 10.2 percent for surface-modified CGS. Some investors in solar technology worry that production of CIGS cells will be limited by the availability of indium. Producing 2 GW of CIGS cells (roughly the amount of silicon cells produced in 2006) would use about 10 percent of the indium produced in 2004. For comparison, silicon solar cells used up 33 percent of the world's electronic grade silicon production in 2006. Nano solar claims to waste only 5 percent of the indium it uses. As of 2006, the best conversion efficiency for flexible CIGS cells on polyimide is 14.1 percent by Tiwari et al., at the ETH, Switzerland. Comparable efficiencies have been reported on other flexible substrates.

That being said, indium can easily be recycled from decommissioned PV modules. The recycling program in Germany is an example that highlights the regenerative industrial paradigm: "From cradle to cradle."

Se allows for better uniformity across the layer and so the number of recombination sites in the film are reduced which benefits the quantum efficiency and thus the conversion efficiency.

Gallium Arsenide (GaAs) Multi Junction

High-efficiency multi-junction cells were originally developed for special applications such as satellites and space exploration, but at present, their use in terrestrial concentrators might be the lowest cost alternative in terms of \$/kWh and \$/W. These multi-junction cells consist of multiple thin films produced using Metalorganic vapor phase epitaxy. A triple-junction cell for example, may consist of the semiconductors: GaAs, Ge, and $GaInP_2$. Each type of semiconductor will have a characteristic band gap energy which, loosely speaking, causes it to absorb light most efficiently at a certain color, or more precisely, to absorb electromagnetic radiation over a portion of the spectrum. The semiconductors are carefully chosen to absorb nearly all of the solar spectrum, thus generating electricity from as much of the solar energy as possible.

This technology was utilized in the Mars Rover missions.

Tandem solar cells based on monolithic, series connected, gallium indium phosphide (GaInP), gallium arsenide GaAs, and germanium Ge p-n junctions, are seeing demand rapidly rise. In just the past 12 months (12/2006 - 12/2007), the cost of 4N gallium metal has risen from about $350 per kg to $680 per kg. Additionally, germanium metal prices have risen substantially to $1000-$1200 per kg this year. Those materials include gallium (4N, 6N and 7N Ga), arsenic (4N, 6N and 7N) and germanium, pyrolitic boron nitride (pBN) crucibles for growing crystals, and boron oxide, these products are critical to the entire substrate manufacturing industry.

Triple-junction GaAs solar cells were also being used as the power source of the Dutch four-time World Solar Challenge winners Nuna in 2005 and 2007, and also by the Dutch solar cars Solutra and Twente One.

The Dutch Radboud University Nijmegen set the record for thin film solar cell effiency using a single junction GaAs to 25.8% in August 2008 using only 4 μm thick GaAs layer which can be transferred from a wafer base to glass or plastic film.

Light-absorbing Dyes (DSSC)

Typically, a ruthenium metalorganic dye (Ru-centered) is used as a monolayer of light-absorbing material. The dye-sensitized solar cell depends on a mesoporous layer of nanoparticulate titanium dioxide to greatly amplify the surface area (200-300 m^2/g TiO_2, as compared to approximately 10 m^2/g of flat single crystal). The photogenerated electrons from the *light absorbing dye* are passed on to the *n-type* TiO_2, and the holes are passed to an electrolyte on the other side of the dye. The circuit is completed by a redox couple in the electrolyte, which can be liquid or solid. This type of cell allows a more flexible use of materials, and is typically manufactured by screen printing, with the potential for lower processing costs than those used for *bulk* solar cells. However, the dyes in these cells also suffer from degradation under heat and UV light, and the cell casing is difficult to seal due to the solvents used in assembly. In spite of the above, this is a popular emerging technology with some commercial impact forecast within this decade.

Organic/Polymer Solar Cells

Organic solar cells and polymer solar cells are built from thin films (typically 100 nm) of organic semiconductors such as polymers and small-molecule compounds like polyphenylene vinylene, copper phthalocyanine (a blue or green organic pigment) and carbon fullerenesand fullerene derivatives such as PCBM. Energy conversion efficiencies achieved to date using conductive polymers are low compared to inorganic materials, with the highest reported efficiency of 6.5 percent for a tandem cell architecture. However, these cells could be beneficial for some applications where mechanical flexibility and disposability are important.

These devices differ from inorganic semiconductor solar cells in that they do not rely on the large built-in electric field of a p-n junction to separate the electrons and holes created when photons are absorbed. The active region of an organic device consists of two materials, one which acts as an electron donor and the other as an acceptor. When a photon is converted into an electron hole pair, typically in the donor material, the charges tend to remain bound in the form of an exciton, and are separated when the exciton diffuses to the donor-acceptor interface. The short exciton diffusion lengths of most polymer systems tend to limit the efficiency of such devices. Nanostructured interfaces, sometimes in the form of bulk heterojunctions, can improve performance.

Silicon Thin Films

Silicon thin-film cells are mainly deposited by chemical vapor deposition (typically plasma-enhanced (PE-CVD)) from silane gas and hydrogen gas. Depending on the deposition parameters, this can yield:

1. Amorphous silicon (a-Si or a-Si:H),

2. Protocrystalline silicon,

3. Nanocrystalline silicon (nc-Si or nc-Si:H), also called microcrystalline silicon.

These types of silicon present dangling and twisted bonds, which results in deep defects (energy levels in the band gap) as well as deformation of the valence and conduction bands (band tails). The solar cells made from these materials tend to have lower *energy conversion efficiency* than *bulk* silicon, but are also less expensive to produce. The quantum efficiency of thin film solar cells is also lower due to reduced number of collected charge carriers per incident photon.

Amorphous silicon has a higher band gap (1.7 eV) than crystalline silicon (c-Si) (1.1 eV), which means it absorbs the visible part of the solar spectrum more strongly than the infrared portion of the spectrum. As nc-Si has about the same band gap as c-Si, the nc-Si and a-Si can advantageously be combined in thin layers, creating a layered cell called a tandem cell. The top cell in a-Si absorbs the visible light and leaves the infrared part of the spectrum for the bottom cell in nanocrystalline Si.

Recently, solutions to overcome the limitations of thin-film crystalline silicon have been developed. Light trapping schemes where the weakly absorbed long wavelength light is obliquely coupled into the silicon and traverses the film several times can significantly enhance the absorption of sunlight in the thin silicon films. Thermal processing techniques can significantly enhance the crystal quality of the silicon and thereby lead to higher efficiencies of the final solar cells.

A silicon thin film technology is being developed for building integrated photovoltaics (BIPV) in the form of semi-transparent solar cells which can be applied as window glazing. These cells function as window tinting while generating electricity.

Nanocrystalline Solar Cells

These structures make use of some of the same thin-film light absorbing materials but are overlain as an extremely thin absorber on a supporting matrix of conductive polymer or mesoporous metal oxide having a very high surface area to increase internal reflections (and hence increase the probability of light absorption). Using nanocrystals allows one to design architectures on the length scale of nanometers, the typical exciton diffusion length. In particular, single-nanocrystal ('channel') devices, an array of single p-n junctions between the electrodes and separated by a period of about a diffusion length, represent a new architecture for solar cells and potentially high efficiency.

Concentrating Photovoltaics (CPV)

Concentrating photovoltaic systems use a large area of lenses or mirrors to focus sunlight on a small area of photovoltaic cells. If these systems use single or dual-axis tracking to improve performance, they may be referred to as *Heliostat Concentrator Photovoltaics* (HCPV). The primary attraction of CPV systems is their reduced usage of semiconducting material which is expensive and currently in short supply. Additionally, increasing the concentration ratio improves the performance of general photovoltaic materials. Despite the advantages of CPV technologies their application has been limited by the costs of focusing, tracking and cooling equipment. On October 25, 2006, the Australian federal government and the Victorian state government together with photovoltaic technology company Solar Systems announced a project using this technology, solar power station in Victoria, planned to come online in 2008 and be completed by 2013.

Silicon Solar Cell Device Manufacture

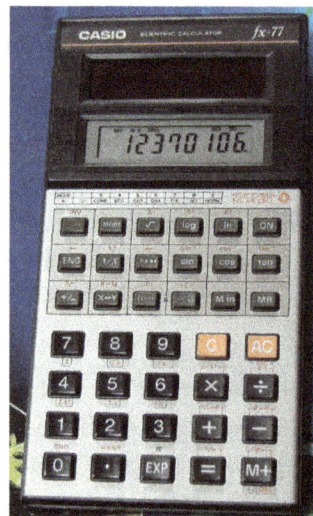

Solar powered scientific calculator

Because solar cells are semiconductor devices, they share many of the same processing and manufacturing techniques as other semiconductor devices such as computer and memory chips. However, the stringent requirements for cleanliness and quality control of semiconductor fabrication are a little more relaxed for solar cells. Most large-scale commercial solar cell factories today make screen printed poly-crystalline silicon solar cells. Single crystalline wafers which are used in the semiconductor industry can be made into excellent high efficiency solar cells, but they are generally considered to be too expensive for large-scale mass production.

Poly-crystalline silicon wafers are made by wire-sawing block-cast silicon ingots into very thin (180 to 350 micrometer) slices or wafers. The wafers are usually lightly p-type doped. To make a solar cell from the wafer, a surface diffusion of n-type dopants is performed on the front side of the wafer. This forms a p-n junction a few hundred nanometers below the surface.

Antireflection coatings, which increase the amount of light coupled into the solar cell, are typically next applied. Over the past decade, silicon nitride has gradually replaced titanium dioxide as the antireflection coating of choice because of its excellent surface passivation qualities (i.e., it prevents carrier recombination at the surface of the solar cell). It is typically applied in a layer several hundred nanometers thick using plasma-enhanced chemical vapor deposition (PECVD). Some solar cells have textured front surfaces that, like antireflection coatings, serve to increase the amount of light coupled into the cell. Such surfaces can usually only be formed on single-crystal silicon, though in recent years methods of forming them on multicrystalline silicon have been developed.

The wafer then has a full area metal contact made on the back surface, and a grid-like metal contact made up of fine "fingers" and larger "busbars" are screen-printed onto the front surface using a silver paste. The rear contact is also formed by screen-printing a metal paste, typically aluminum. Usually this contact covers the entire rear side of the cell, though in some cell designs it is printed in a grid pattern. The paste is then fired at several hundred degrees Celsius to form metal electrodes in ohmic contact with the silicon. After the metal contacts are made, the solar cells are interconnected in series (and/or parallel) by flat wires or metal ribbons, and assembled into modules or "solar panels." Solar panels have a sheet of tempered glass on the front, and a polymer encapsulation on the back. Tempered glass cannot be used with amorphous silicon cells because of the high temperatures during the deposition process.

Lifespan

Most commercially available solar cells are capable of producing electricity for at least twenty years without a significant decrease in efficiency.

Costs

Cost is established in cost-per-watt and in cost-per-watt in 24 hours for infrared capable photovoltaic cells.

Slicing Costs

University of Utah engineers devised a new way to slice thin wafers of the chemical element germanium for use in the most efficient type of solar power cells. The new method should lower the cost of such cells by reducing the waste and breakage of the brittle semiconductor.

Low Cost Solar Cell

Dye-sensitized solar cell is considered the low-cost solar cell.

This cell is extremely promising because it is made of low-cost materials and does not need elaborate apparatus to manufacture, so it can be made in a DIY way allowing more players to produce it than any other type of solar cell. In bulk it should be significantly less expensive than older solid-state cell designs. It can be engineered into flexible sheets. Although its conversion efficiency is less than the best thin film cells, its price/performance ratio should be high enough to allow them to compete with fossil fuel electrical generation.

Current Research on Materials and Devices

There are currently many research groups active in the field of photovoltaics in universities and research institutions around the world. This research can be divided into three areas: making current technology solar cells cheaper and/or more efficient to effectively compete with other energy sources; developing new technologies based on new solar cell architectural designs; and developing new materials to serve as light absorbers and charge carriers.

Silicon Processing

One way of reducing the cost is to develop cheaper methods of obtaining silicon that is sufficiently pure. Silicon is a very common element, but is normally bound in silica, or silica sand. Processing silica (SiO_2) to produce silicon is a very high energy process - at current efficiencies, it takes over two years for a conventional solar cell to generate as much energy as was used to make the silicon it contains. More energy efficient methods of synthesis are not only beneficial to the solar industry, but also to industries surrounding silicon technology as a whole.

The current industrial production of silicon is via the reaction between carbon (charcoal) and silica at a temperature around 1700 degrees Celsius. In this process, known as carbothermic reduction, each tonne of silicon (metallurgical grade, about 98 percent pure) is produced with the emission of about 1.5 tonnes of carbon dioxide.

Solid silica can be directly converted (reduced) to pure silicon by electrolysis in a molten salt bath at a fairly mild temperature (800 to 900 degrees Celsius). While this new process is in principle the same as the FFC Cambridge Process which was first discovered in late 1996, the interesting laboratory finding is that such electrolytic silicon is in the form of porous silicon which turns readily into a fine powder, (with a particle size of a few micrometers), and may therefore offer new opportunities for development of solar cell technologies.

Another approach is also to reduce the amount of silicon used and thus cost, is by micromachining wafers into very thin, virtually transparent layers that could be used as transparent architectural coverings. The technique involves taking a silicon wafer, typically 1 to 2 mm thick, and making a multitude of parallel, transverse slices across the wafer, creating a large number of slivers that have a thickness of 50 micrometers and a width equal to the thickness of the original wafer. These slices are rotated 90 degrees, so that the surfaces corresponding to the faces of the original wafer become the edges of the slivers. The result is to convert, for example, a 150 mm diameter, 2 mm-thick wafer having an exposed silicon surface area of about 175 cm² per side into about 1000 slivers having dimensions of 100 mm × 2 mm × 0.1 mm, yielding a total exposed silicon surface area of about 2000 cm² per side. As a result of this rotation, the electrical doping and contacts that were on the face of the wafer are located the edges of the sliver, rather than the front and rear as is the case with conventional wafer cells. This has the interesting effect of making the cell sensitive from both the front and rear of the cell (a property known as bifaciality). Using this technique, one silicon wafer is enough to build a 140 watt panel, compared to about 60 wafers needed for conventional modules of same power output.

Thin-film Processing

Thin-film photovoltaic cells can use less than 1 percent of the expensive raw material (silicon or other light absorbers) compared to wafer based solar cells, leading to a significant price drop per Watt peak capacity. There are many research groups around the world actively researching different thin-film approaches and/or materials. However, it remains to be seen if these solutions can achieve a similar market penetration as traditional bulk silicon solar modules.

One particularly promising technology is crystalline silicon thin films on glass substrates. This technology combines the advantages of crystalline silicon as a solar cell material (abundance, non-toxicity, high efficiency, long-term stability) with the cost savings of using a thin-film approach.

Another interesting aspect of thin-film solar cells is the possibility to deposit the cells on all kind of materials, including flexible substrates (PET for example), which opens a new dimension for new applications.

Metamorphic Multi-junction Solar Cell

The ultra-light, highly efficient solar cell was developed and is being commercialized by Emcore Corp. of Albuquerque, New Mexico, in partnership with the Air Force Research Laboratories Space Vehicles Directorate at Kirtland Air Force Base in Albuquerque.

It represents a new class of solar cells with clear advantages in performance, engineering design, operation and cost. For decades, conventional cells have featured wafers of semiconducting materials with similar crystalline structure. Their performance and cost effectiveness is constrained by growing the cells in an upright configuration. Meanwhile, the cells are rigid, heavy and thick with a bottom layer made of germanium.

In the new method, the cell is grown upside down. These layers use high-energy materials with extremely high quality crystals, especially in the upper layers of the cell where most of the power is produced. Not all of the layers follow the lattice pattern of even atomic spacing. Instead, the cell includes a full range of atomic spacing, which allows for greater absorption and use of sunlight. The thick, rigid germanium layer is removed, reducing the cell's cost and 94 percent of its weight. By turning the conventional approach to cells on its head, the result is an ultra-light and flexible cell that also converts solar energy with record efficiency (40.8 percent under 326 suns concentration).

Polymer Processing

The invention of conductive polymers (for which Alan Heeger, Alan G. MacDiarmid and Hideki Shirakawa were awarded a Nobel prize) may lead to the development of much cheaper cells that are based on inexpensive plastics. However, all organic solar cells made to date suffer from degradation upon exposure to UV light, and hence have lifetimes which are far too short to be viable. The conjugated double bond systems in the polymers, which carry the charge, are always susceptible to breaking up when radiated with shorter wavelengths. Additionally, most conductive polymers, being highly unsaturated and reactive, are highly sensitive to atmospheric moisture and oxidation, making commercial applications difficult.

Nanoparticle Processing

Experimental non-silicon solar panels can be made of quantum heterostructures, eg. carbon nanotubes or quantum dots, embedded in conductive polymers or mesoporous metal oxides. In addition, thin films of many of these materials on conventional silicon solar cells can increase the optical coupling efficiency into the silicon cell, thus boosting the overall efficiency. By varying the size of the quantum dots, the cells can be tuned to absorb different wavelengths.

Researchers located at the University of California, San Diego have come up with a way of making solar photovoltaic cells more efficient by making them fuzzy with indium phosphide nanowires.

Transparent Conductors

Many new solar cells use transparent thin films that are also conductors of electrical charge. The dominant conductive thin films used in research now are transparent conductive oxides ("TCO"), and include fluorine-doped tin oxide (SnO_2:F, or "FTO"), doped zinc oxide (e.g.: ZnO:Al), and indium tin oxide ("ITO"). These conductive films are also used in the LCD industry for flat panel displays. The dual function of a TCO allows light to pass through a substrate window to the active light absorbing material beneath, and also serves as an ohmic contact to transport photogenerated charge carriers away from that light absorbing material. The present TCO materials are effective for research, but perhaps are not yet optimized for large-scale photovoltaic production. They require very special deposition conditions at high vacuum, they can sometimes suffer from poor mechanical strength, and most have poor transmittance in the infrared portion of the spectrum (e.g.: ITO thin films can also be used as infrared filters in airplane windows). These factors make large-scale manufacturing more costly.

A relatively new area has emerged using carbon nanotube networks as a transparent conductor for organic solar cells. Nanotube networks are flexible and can be deposited on surfaces a variety of ways. With some treatment, nanotube films can be highly transparent in the infrared, possibly enabling efficient low band gap solar cells. Nanotube networks are p-type conductors, whereas traditional transparent conductors are exclusively n-type. The availability of a p-type transparent conductor could lead to new cell designs that simplify manufacturing and improve efficiency.

Silicon Wafer based Solar Cells

Despite the numerous attempts at making better solar cells by using new and exotic materials, the reality is that the photovoltaics market is still dominated by silicon wafer-based solar cells (first-generation solar cells). This means that most solar cell manufacturers are currently equipped to produce this type of solar cells. Consequently, a large body of research is being done all over the world to manufacture silicon wafer-based solar cells at lower cost and to increase the conversion efficiencies without an exorbitant increase in production cost. The ultimate goal for both wafer based and alternative photovoltaic concepts is to produce solar electricity at a cost comparable to currently marked dominating technologies like coal and nuclear power in order to make it the leading primary energy source. To achieve this it may be necessary to reduce the cost of installed solar systems from currently about US$ 1.80 (for bulk Si technologies) to about US$ 0.50 per Watt peak power. Since a major part of the final cost of a traditional bulk silicon module is related to the high cost of solar grade poly silicon feedstock (about US$ 0.4/Watt peak) there exists substantial drive to make Si solar cells thinner (material savings) or to make solar cells from cheaper upgraded metallurgical silicon (so called "dirty Si").

IBM has a semiconductor wafer reclamation process that uses a specialized pattern removal technique to repurpose scrap semiconductor wafers to a form used to manufacture silicon-based solar panels. The new process was recently awarded the "2007 Most Valuable Pollution Prevention Award" from The National Pollution Prevention Roundtable (NPPR).

Infrared Solar Cells

Researchers at Idaho National Laboratory, along with partners at Microcontinuum Inc. (Cambridge, MA) and Patrick Pinhero of the University of Missouri, have devised an inexpensive way to produce plastic sheets containing billions of nanoantennas that collect heat energy generated by the sun and other sources, which garnered two 2007 Nano50 awards. The technology is the first step toward a solar energy collector that could be mass-produced on flexible materials. While methods to convert the energy into usable electricity still need to be developed, the sheets could one day be manufactured as lightweight "skins" that power everything from hybrid cars to computers and iPods with higher efficiency than traditional solar cells. The nanoantennas also have the potential to act as cooling devices that draw waste heat from buildings or electronics without using electricity. The nanoantennas target mid-infrared rays, which the Earth continuously radiates as heat after absorbing energy from the sun during the day; also double-sided nanoantenna sheets can harvest energy from different parts of the sun's spectrum. In contrast, traditional solar cells can only use visible light, rendering them idle after dark.

Also, Konarka is researching infrared light activated photovoltaics which would enable night-time power generation.

UV Solar Cells

Japan's National Institute of Advanced Industrial Science and Technology (AIST) has succeeded in developing a transparent solar cell that uses ultraviolet light to generate electricity but allows visible light to pass through it. Most conventional solar cells use visible and infrared light to generate electricity. In contrast, the innovative new solar cell uses ultraviolet radiation. used to replace conventional window glass, the installation surface area could be large, leading to potential uses that take advantage of the combined functions of power generation, lighting and temperature control.

3-D Solar Cells

Three-dimensional solar cells that capture nearly all of the light that strikes them and could boost the efficiency of photovoltaic (PV) systems while reducing their size, weight and mechanical complexity. The new 3D solar cells capture photons from sunlight using an array of miniature "tower" structures that resemble high-rise buildings in a city street grid.

Antireflective and All-angle Coating

With a reflective coating, developed at RPI, the material absorbs 96.21 per cent of incident sunlight. This gain in absorption was consistent across the entire spectrum of sunlight, from UV to visible light and infrared.

Such a method tackles the problem of absorbing sunlight evenly and equally from all angles. The seven layers, each with a height of 50-100 nm, are made up of silicon dioxide and titanium dioxide nanorods positioned at an oblique angle. Each layer looks and functions similar to a dense forest where sunlight is "captured" between the trees. The nanorods were attached to a silicon substrate via chemical vapor disposition and the new coating can be affixed to nearly any photovoltaic materials for use in solar cells, including III-V multi-junction and cadmium telluride.

Metamaterials

Researchers at Duke University and Boston College have engineered a metamaterial that utilizes tiny geometric shapes to absorb both the electrical and magnetic properties of electromagnetic waves over a certain frequency range at a level that meets standards of scientific perfection. This results in the total absorption of light, turning it into heat, which can then create energy.

Validation, Certification and Manufacturers

National Renewable Energy Laboratory tests and validates solar technologies. There are three reliable certifications of solar equipment: UL and IEEE (both U.S. standards) and IEC.

Solar cells are manufactured primarily in Japan, China, Germany, Taiwan, and the USA, though numerous other nations have or are acquiring significant solar cell production capacity. While technologies are constantly evolving toward higher efficiencies, the most effective cells for low cost electrical production are not necessarily those with the highest efficiency, but those with a balance between low-cost production and efficiency high enough to minimize area-related balance of systems cost. Those companies with large scale manufacturing technology for coating inexpensive substrates may, in fact, ultimately be the lowest cost net electricity producers, even with cell efficiencies that are lower than those of single-crystal technologies.

Solar Panel

A solar panel is a collection of solar (or photovoltaic) cells, which can be used to generate electricity through photovoltaic effect. These cells are arranged in a grid-like pattern on the surface of solar panels.

Thus, it may also be described as a set of photovoltaic modules, mounted on a structure supporting it. A photovoltaic (PV) module is a packaged and connected assembly of 6×10 solar cells.

When it comes to wear-and-tear, these panels are very hardy. Solar panels wear out extremely slow. In a year, their effectiveness decreases only about one to two per cent (at times, even lesser).

Generations of Solar Panels

The material most often used for solar cells is silicon. This material can occur in three forms:

- Monocrystalline silicon,

- Polycrystalline (or multicrystalline) silicon,

- Amorphous silicon.

These various forms produce different types of solar panels with differing prices, useful lives and output:

- First-generation solar panels use monocrystalline or polycrystalline silicon (output of 12 to 19 %).

- Second-generation solar panels consist of solar cells made of amorphous silicon. This name also applies to solar panels based on other materials that have appeared on the market more recently:

  - CIS (copper-indium-selenium),

  - CIS (copper-indium-gallium-selenium),

  - CdTe (cadmium telluride).

The particular feature of this second generation is that it uses thin semiconductor layers ("thin films"). This explains why these panels are less expensive and more aesthetic, but also has a lower output (from 5 to 11 %).

Size of Solar Panel

These days, most solar panels are about 5 and a half feet tall and a little more than 3 feet wide.

If you look closely at the solar panel in the image above, you'll notice 60 little squares. These squares are actually individual solar "cells," which are linked together by wires. The cells are where electricity is made, and the wires carry the electricity to a junction box where the panel is hooked into a larger array.

Importance of Solar Panel Size

The more solar cells working in tandem, the more power they'll create. That's why the size of the panel matters if you're trying to calculate how much electricity a panel makes.

Solar panels have been about this size for decades, but modern panels make more electricity than in the past. That's because panel manufacturers have found ways to improve cell efficiency over time.

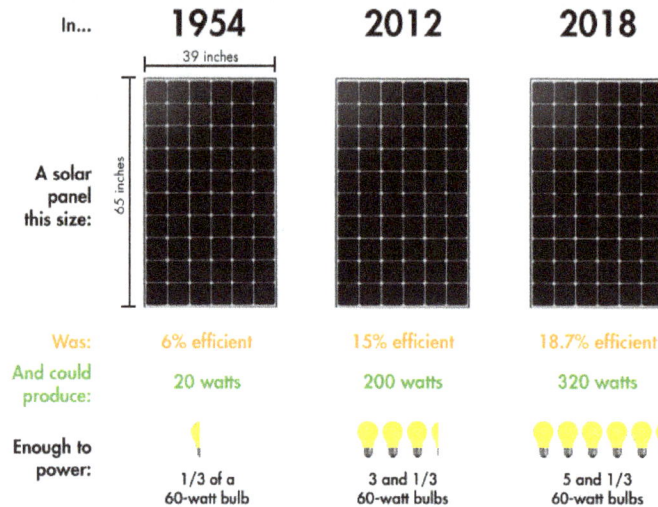

Efficiency of Solar Panels

Solar efficiency relates to the amount of available energy from the sun that gets converted into electricity.

Back in the 1950s, the first solar cells were capable of taking 6% of the energy from the sun and converting it into electricity.

If they were configured to be the same array of 60 cells you see in the image above, that would have created a current of about 20 watts electricity, about a third of what would be needed to light up a 60 watt incandescent bulb.

When we originally wrote this page in 2012, solar cells could convert 15% of the energy hitting them from the sun into power. As of 2018, the efficiency of the most advanced solar cells is closer to 23%, while average solar cells for residential use are around 18.7% efficient.

If you combine the efficiency of the cells with the size of the panel, you get a number called the "power rating." In the solar industry, we say "that panel is rated to produce X watts."

Solar Panel Design

Most solar cells are a few square centimeters in area and protected from the environment by a thin coating of glass or transparent plastic. Because a typical 10 cm × 10 cm (4 inch × 4 inch) solar cell generates only about two watts of electrical power (15 to 20 percent of the energy of light incident on their surface), cells are usually combined in series to boost the voltage or in parallel to increase the current. A solar, or photovoltaic (PV), module generally consists of 36 interconnected cells laminated to glass within an aluminum frame. In turn, one or more of these modules may be wired and framed together to form a solar panel. Solar panels are slightly less efficient at energy conversion per surface area than individual cells, because of inevitable inactive areas in the assembly and cell-to-cell variations in performance. The back of each solar panel is equipped with standardized sockets so that its output can be combined with other solar panels to form a solar array. A complete photovoltaic system may consist of many solar panels, a power system for accommodating different electrical loads, an external circuit, and storage batteries. Photovoltaic systems are broadly classifiable as either stand-alone or grid-connected systems.

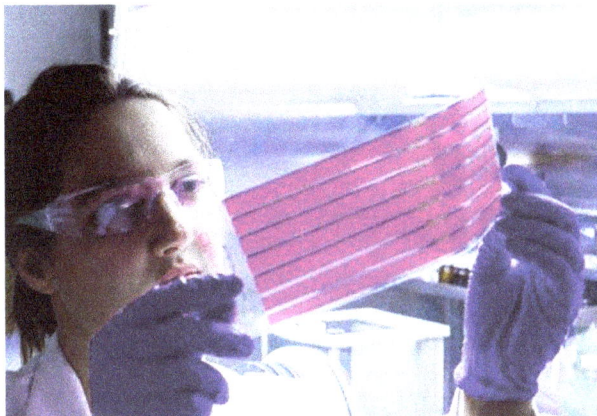

Solar cell a scientist examines a sheet of polymer solar cells, which are more lightweight, more flexible, and cheaper than traditional silicon solar cells.

Stand-alone systems contain a solar array and a bank of batteries directly wired to an application or load circuit. A battery system is essential to compensate for the absence of any electrical output from the cells at night or in overcast conditions; this adds considerably to the overall cost. Each battery stores direct current (DC) electricity at a fixed voltage determined by the panel specifications, although load requirements may differ. DC-to-DC converters are used to provide the voltage levels demanded by DC loads, and DC-to-AC inverters supply power to alternating current (AC) loads. Stand-alone systems are ideally suited for remote installations where linking to a central power station is prohibitively expensive. Examples include pumping water for feedstock and providing electric power to lighthouses, telecommunications repeater stations, and mountain lodges.

Grid-connected systems integrate solar arrays with public utility power grids in two ways. One-way systems are used by utilities to supplement power grids during midday peak usage. Bidirectional systems are used by companies and individuals to supply some or all of their power needs, with any excess power fed back into a utility power grid. A major advantage of grid-connected systems is that no storage batteries are needed. The corresponding reduction in capital and maintenance costs is offset, however, by the increased complexity of the system. Inverters and additional protective gear are needed to interface low-voltage DC output from the solar array with a high-voltage AC power grid. Additionally, rate structures for reverse metering are necessary when residential and industrial solar systems feed energy back into a utility grid.

The simplest deployment of solar panels is on a tilted support frame or rack known as a fixed mount. For maximum efficiency, a fixed mount should face south in the Northern Hemisphere or north in the Southern Hemisphere, and it should have a tilt angle from horizontal of about 15 degrees less than the local latitude in summer and 25 degrees more than the local latitude in winter. More complicated deployments involve motor-driven tracking systems that continually reorient the panels to follow the daily and seasonal movements of the Sun. Such systems are justified only for large-scale utility generation using high-efficiency concentrator solar cells with lenses or parabolic mirrors that can intensify solar radiation a hundredfold or more.

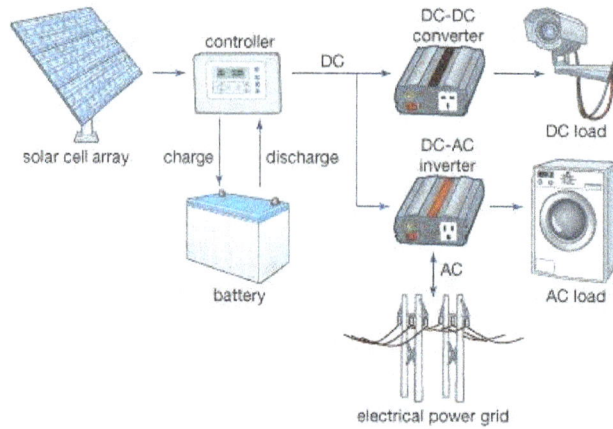

A grid-connected solar cell system.

Although sunlight is free, the cost of materials and available space must be considered in designing a solar system; less-efficient solar panels imply more panels, occupying more space, in order to produce the same amount of electricity. Compromises between cost of materials and efficiency are particularly evident for space-based solar systems. Panels used on satellites have to be extra-rugged, reliable, and resistant to radiation damage encountered in Earth's upper atmosphere. In addition, minimizing the liftoff weight of these panels is more critical than fabrication costs. Another factor in solar panel design is the ability to fabricate cells in "thin-film" form on a variety of substrates, such as glass, ceramic, and plastic, for more flexible deployment. Amorphous silicon is very attractive from this viewpoint. In particular, amorphous silicon-coated roof tiles and other photovoltaic materials have been introduced in architectural design and for recreational vehicles, boats, and automobiles.

Thin-film solar cell Thin-film solar cells, such as those used in solar panels, convert light energy into electrical energy.

Solar Panels Work

Solar panels collect clean renewable energy in the form of sunlight and convert that light into electricity which can then be used to provide power for electrical loads. Solar panels are comprised of several individual solar cells which are themselves composed of layers of silicon, phosphorous (which provides the negative charge), and boron (which provides the positive charge). Solar panels absorb the photons and in doing so initiate an electric current. The resulting energy generated from photons striking the surface of the solar panel allows electrons to be knocked out of their atomic orbits and released into the electric field generated by the solar cells which then pull these free electrons into a directional current. This entire process is known as the Photovoltaic effect. An average home has more than enough roof area for the necessary number of solar panels to produce enough solar electricity to supply all of its power needs excess electricity generated goes onto the main power grid, paying off in electricity use at night.

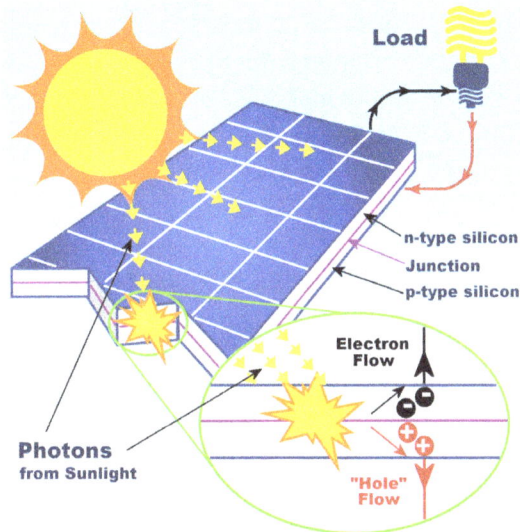

In a well-balanced grid-connected configuration, a solar array generates power during the day that is then used in the home at night. In off-grid solar applications, a battery bank, charge controller, and in most cases, an inverter are necessary components. The solar array sends direct current (DC) electricity through the charge controller to the battery bank. The power is then drawn from the battery bank to the inverter, which converts the DC current into alternating current (AC) that can be used for non-DC appliances. Assisted by an inverter, solar panel arrays can be sized to meet the most demanding electrical load requirements. The AC current can be used to power loads in homes or commercial buildings, recreational vehicles and boats, remote cabins, cottages, or homes, remote traffic controls, telecommunications equipment, oil and gas flow monitoring, RTU, SCADA, and much more.

Cons of Solar Panels

1. Unreliable

   Despite the promise of electricity of solar panels, critics say that these are not reliable sources of energy or electricity supply since they cannot operate at a maximum under some weather conditions like storms, cloudy and rainy days. Since they are dependent on the light coming from the sun, it is not as reliable during rainy weather and in months when not enough solar energy coming from the sun is available. This is on top of the solar panels not able to store energy at night.

2. Expensive

    Homeowners who might want to use solar panels need to invest a fairly high amount of money not only for the purchase of these panels but also for installation. Money should be presented upfront and payments should be made for the inverter, batteries, panels and wiring. These expenses are costly. Moreover, if the homeowner chooses to install solar panels, conversion and remodeling of the house might be needed which can also be expensive.

3. Pollution

    Although using solar panels is associated to green living and the pollution they emit is not as much as chemicals and toxic wasted from other sources of energy such as coal and fossil, it is said that the installation and transportation of solar panels contribute to the emission of green gases in the atmosphere. Moreover, manufacturing of these solar panels make use of hazardous products and toxic materials.

4. Costs of Energy Storage

    Apart from the expenses incurred for the acquisition and installation of solar panels, there is a need to store energy that will be used by the solar systems. If energy collected is not utilized right away, storing them in batteries is imperative. For these batteries to be operational in solar systems that are in remote areas with no commercial electricity, they need to be charged during the day. This process is expensive since large batteries need to be available especially if there is no electricity available during the night or the whole day.

5. Space Requirement

    If a property or structure needs to get more electricity than usual, numerous solar panels should be installed and larger ones should be used. This number of solar systems can take a lot of space especially if the roof is not big enough to accommodate the panels. In order to solve the problem, there should be extra space for installation and this space should be accessible by sunlight.

Solar Shingle

Solar shingles on a home in San Diego, Calif

Solar shingles are an integrated photovoltaic (PV) building product, which means they directly generate electricity from sunlight. They're made with specific materials that naturally undergo an electronic process in the presence of sunlight.

From a distance, they look like ordinary roof shingles. But up close, it's obvious they're more than that. Solar shingles are used on the roofs of commercial and residential buildings to generate electricity, and some models can convert the electricity to heat.

Solar shingles are made of the same materials as those used in regular solar panels, including wire, a photosensitive waterproof product to protect the roof from outdoor elements, and material that can generate an electric current in sunlight.

In fact, solar shingles and solar panels operate the same way and are overall quite similar. The main difference between them is cost. Solar panels have one function: They generate electricity. But solar shingles serve a dual purpose - they generate electricity and act as shingles, which makes them less expensive in the long run.

Types of Solar Shingle

There are currently two basic types of photovoltaic solar shingles used to produce home solar power. Traditional silicon solar cells and the newly developed copper-indium-gallium-selenide (CIGS) thin film photovoltaics.

- Silicon Solar Shingles

   Silicon solar shingles are very much like a traditional solar panel. They are rigid and in a frame yet made thin and proportional to conventional roofing materials and are installed in line with existing roofing materials.

- CIGS Solar Shingles

   CIGS photovoltaic roof shingles utilize the newest thin film technology for producing home solar power. CIGS are able to be made into more flexible and lighter weight shingles that are able to blend into an existing or new roof.

Working of Solar Shingle

A photon is a unit of electromagnetic radiation with no charge; Its protons (positively charged subatomic particles) and electrons (negatively charged subatomic particles) are equal, giving it no charge. The photon, here, is a particle of sunlight. Sunlight strikes a solar shingle coated with a special surface that naturally knocks off an electron from the particle of light. The freed electron travels through an electrical circuit to an area where other electrons are stored. This collection of electrons is then harvested to generate a current. This is how all solar electricity works, whether it's a solar shingle, solar panel or something else. Photovoltaic (PV) devices like these can be used to power anything from a calculator to a skyscraper.

Solar shingles, in theory, can produce 100 percent of a building's electricity usage, but it depends on a number of factors: the demand of a building, the amount of sunlight the structure gets at that particular geographic location, the building's current utility rates and availability of sufficient space on the roof that opens to southern skies.

Solar shingles can work practically anywhere (even in gray-weather places) because they can use diffused, scattered sunshine on overcast or even rainy days.

Installing Solar Shingles

Solar shingles are about the same size as a typical roof shingle, while solar panels are flat, rigid panes often several square feet. It's because of this size difference that installation of solar shingles is much more labor-intensive -- more shingles are needed to complete a job compared to the number of panels required for the same project. Installers tediously put in one shingle after another after another and wire each one inside the building.

Photovoltaic (PV) installations cost between $9 and $11 per watt. An average system will range from $15,000 to $30,000 (after incentives). Price is affected by the size of your PV system and the amount of sun and shading in your location.

The time it takes to do an installation varies and is affected by roof space, the number of people in an installation crew and the status of the building (re-roofing verses new construction). Once all the solar shingles are put in and wired together indoors, under the roof, the installer applies for a permit to connect your system to the electric grid.

Needless to say, homeowners can't (or shouldn't) attempt installation on their own. However, the industry is changing, and five to 10 years from now, homeowners may be able to safely and easily "plug and play," according to the American Solar Energy Society (ASES). For this to happen, more sophisticated infrastructure would have to be in place, specifically, smart meters and a smart grid, which don't yet exist. Here's how it would work: The homeowner would plug his or her solar shingles into a smart meter - a meter with plug-in ports for solar electric applications like solar shingles. Through this special meter, they could hook up to the electric grid.

Benefits of Solar Shingle

There are many benefits of using solar shingles and these are some of the major ones:

- They can be integrated into any exterior design of your home.
- They look regular roofing shingles.
- It eliminates the necessity of having large surfaces for traditional solar panel installation.
- Aesthetically-pleasing solar material.
- Solar shingles come in different sizes and models; solid shingle-sized panels, semi-rigid silicon cells, and systems with a thin solar film.
- Easy installation can be done stapling the material onto the roof.
- You can qualify for tax incentives while installing solar shingles, depending on your state policy.
- It can be combined with conventional asphalt shingles.
- Solar shingles, if installed correctly, can withstand high winds.

- As a newer material, they are more effective in comparison with traditional panels.

- Can cover larger areas when compared to traditional solar panels.

- Some photovoltaic shingles can produce around 13-17 watts of electricity.

Drawbacks of Solar Shingles

As any product, solar shingles also have some disadvantages like:

- No energy storage. There is no energy storage on these systems, and you will have to rely on the power grid during storms or at night.

- Your roof has to be at a certain angle to catch the sun. For the solar shingles to catch the sun's rays enough to generate electricity, your roof has to be at the right angle with enough surface area to power your building.

- If possible install the shingles during the construction or remodeling process of your unit.

- The installation cost of solar shingles can be steep.

Balance of System

A Solar PV balance-of-system or BOS refers to the components and equipment that move DC energy produced by solar panels through the conversion system which in turn produces AC electricity.

Most often, BOS refers to all components of a PV system other than the modules. In addition to inverters and racking, this includes the cables/wires, switches, enclosures, fuses, ground fault detectors, and more. BOS applies to all types of solar applications.

BOS components include the majority of the pieces, which make up roughly 10%-50% of solar purchasing and installation costs and account for the majority of maintenance requirements. Essentially it is through the balance-of-system components those we: control cost, increase efficiency, and modernize solar PV systems.

Balance-of-systems cost and efficiency can enable business growth by way of:

a. Expansion into new sectors (i.e. Government, Schools, Non-profit, Agricultural)

b. Word of mouth recommendations

c. Creating new jobs and business locations

d. Increase training and development resources

Balance-of-Systems Efficiency

How customized planning can lead to a cost-effective and energy-efficient BOS environment.

What to look for:

1. Solar Mapping to target solar power or photovoltaic potential per day, by location.

2. Review utility bills with a prospective customer to help them better understand energy usage and fluctuations as well as how much they pay per kilowatt-hour.

3. Inspect and recommend insulation to increase energy efficiency and rebate opportunities.

4. Estimate and compare costs for rooftop and ground-mounted solar PV panels, based on optimum southern exposure.

5. Calculate ways to install a solar system to determine which balance-of-systems components will be optional and which will be necessary.

6. Show short-term value by adding monthly utility savings to qualifying rebates/incentives and subtracting the sum from solar PV system installation cost.

7. Rely on experience to recommend low maintenance and seamless BOS operations.

Solar Inverter

Solar Inverter is a device capable to convert DC into AC electricity. Inverters are typical components of solar electric systems since solar panels generate DC electricity and most devices used in homes or offices operate on AC voltage.

Types of Solar Inverters

There are different types of inverters available for solar power systems, such as off-grid inverters, grid-tie inverters, pure sine wave inverters, and modified sine wave inverters. Ultimately, they all have the same end goal of converting DC solar energy into AC energy that can power your appliance, but the way in which they function differs slightly.

Grid-tied Inverter

Grid-tied inverter is the heart of any grid-tied solar system since a grid-tied system must contain an inverter. A grid-tied inverter converts the DC voltage from the solar array into AC voltage that can be either used right away or exported to the utility grid. A grid-tied inverter must strictly comply with the utility grid's requirements and regulations. For example, grid-tied inverters must generate AC voltage of a strictly sinusoidal form. One of the main features of a grid-tied inverter is that it stops operating in case of a grid failure. Thus technicians doing any repair works on utility network are prevented from getting an electric shock.

Off-grid Inverters

Off-grid inverters are different from grid-tied inverters. An off-grid solar system might not contain an inverter if DC loads only are to be powered. Since off-grid systems are disconnected from the utility grid, off-grid inverters need not to match utility grid requirements and regulations.

Square Wave Solar Power Inverter

The square wave solar power inverter is the simplest and the least expensive type of inverter available. It is generally not used commercially due to its low quality of output power and very large harmonics. Square wave inverters equipped with thyristor output stages chop and invert (hence the name *Inverter*) the DC input positive power to generate a square wave alternating positive to negative AC output signal that is later filtered to approximate a sine wave and eliminate undesired harmonics. Cheaper square wave inverters may also use push-pull transistor circuits with step-up transformers to produce the required output voltage. Square wave inverters are really only used in small stand alone PV systems that will run simple things like lighting or hand tools with universal motors with no problem – but not much else.

Modified Sine Wave Solar Power Inverter

The modified sine wave solar power inverter also called a *quasi-sine wave* inverter, is basically a modified square wave inverter which produces a square wave output with low harmonic distortion and a small "OFF" time between the positive and negative half cycles as the inverter switches polarity. Modified sine wave inverters are suitable for most types of electrical and electronic loads, and are a popular type of inverter on the consumer market today due to their good conversion efficiency, relatively low cost, and can be used in solar installations where waveform shape is not too important. However, modified sine wave inverters may not allow printers, copiers, light dimmers, rechargeable and variable tools to operate correctly due to the switching action of the inverters output stage. Also some audio amplifiers and radios may produce a low frequency background buzz due to the inverters output switching components.

Sine Wave Solar Power Inverter

The sine wave solar power inverter produces a high quality AC voltage waveform with low total harmonic distortion (THD) similar to what you get from your local electricity or utility company. They are used when there is a need for clean sine wave outputs to operate sensitive devices such as electronic equipment, printers, copiers, and stereos, etc.

Pure or true sine wave power inverters are required as standard by most electrical utility companies as part of a grid connected PV solar system. Sine wave inverters have a much higher cost than the previous types for the same wattage due mainly to their better internal electronic circuits. Their output voltage can be controlled either in square wave mode or in pulse width modulated (PWM) mode.

In PWM pure sine wave circuits, the output voltage and frequency are controlled by varying the duty cycle of the high frequency pulses. Chopped voltage then passes through an output LC low pass filter network to produce a clean sinusoidal output. This allows the output voltage and frequency to be well controlled, ensuring that any AC load within the inverters power limits will operate properly. Pure sine wave inverters are generally not suitable for home solar power as their cost is too expensive and are inefficient, instead "stepped" sine wave inverters are used.

Only the top end high quality inverters actually supply a pure sine wave as their output waveform. Also some of the higher end pure sine wave inverters have solar tracking circuits or maximum power point tracking (MPPT) built-in, in order to optimally operate a motorized solar tracking PV array.

Most sine wave models available on the market produce a variation on the modified sine wave inverter above. Their voltage output is not a pure sinusoid but they can operate almost anything that can be connected to the local grid. When it comes to efficiency, sine wave inverters perform better than pure sinusoidal inverters.

Central Inverters

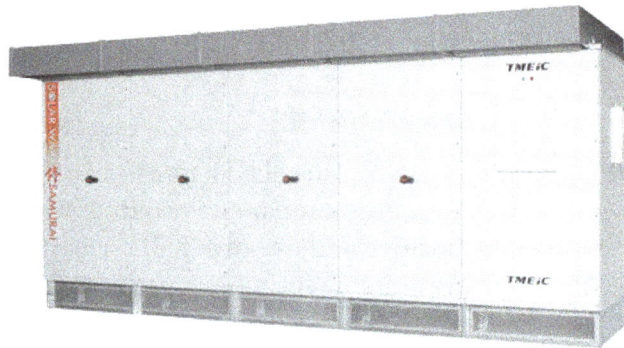

TMEIC central inverter

Central inverters are similar to string inverters but they are much larger and can support more strings of panels. Instead of strings running directly to the inverter, as with string models, the strings are connected together in a common combiner box that runs the DC power to the central inverter where it is converted to AC power. Central inverters require fewer component connections, but require a pad and combiner box. They are best suited for large installations with consistent production across the array.

Battery based Inverter/Chargers

With the growth of solar storage, battery-based inverter/chargers are becoming increasingly important. Battery based inverter/chargers are bi-directional in nature, including both a battery charger and an inverter. They require a battery to operate. Battery-based inverter/chargers may be grid-interactive, standalone grid-tied or off-grid, depending on their UL rating and design. The primary benefit of inverter/chargers is that they provide for continuous operation of critical loads irrespective of the presence or condition of the grid. UL1741 requires the grid-tied generation source to stop generating power in the event of a grid outage. This de-powering is known as anti-islanding, as opposed to 'islanding' which is defined as generating power to power a location in the event of a grid outage. Therefore, UL1741 grid-tie inverters will not generate power in the event of a grid outage, so a user will experience an outage irrespective of the availability solar harvest. Battery-based inverter/chargers will power the critical loads in the event of a grid outage, but will do so in a manner to not create the islanding condition. Further, UL1741 inverter/chargers may be rated as either interactive or standalone. The former export excess power to the grid, while the latter do not—by rating and by definition. In all instances, the battery based inverter/charger manages energy between the array and the grid while keeping the batteries charged. They monitor battery status and regulate how the batteries are charged.

Magnum Energy inverter/charger

Solar Micro-inverter

A micro inverter is as very small inverter designed to be attached directly to one or two solar panels. This is different to a common string solar inverter which is usually located on a wall some distance from the string of solar panels and connected via DC cable; DC power from the string of the panels is then converted to AC at the inverter. Micro's however are attached to each individual panel which means each panel works independently from the rest of the solar array and DC power is converted to AC immediately on the roof.

Advantages of Micro Inverters

1. The core advantage of using micro inverters is that theoretically you can yield more solar electricity. The reason for this is that there are slight differences in voltages between solar panels. When solar panels are in a string the voltage is reduced to the voltage of the lowest voltage panel in the string.

2. If a solar system is facing multiple angles, meaning some panels are facing south, some east, and some west, then micro-inverters are the way to go. Or, if you have shading issues from trees or a large chimney, again micro-inverters would be best. In these situations, the solar panels will be producing different amounts of electricity at different times of the day, but micro-inverters will ensure you harvest all of the energy, while with a standard inverter you will lose some of this production. With solar panels all facing one direction on your system, and you have marginal shading issues, then your best option is a standard inverter. You'll get about the same production, without paying the higher cost.

3. Optimizers are an option for standard inverters as well, which function very similarly to a micro-inverter. With an optimizer, you still have a standard inverter, but you also have optimizers for each individual panel combating production differences.

4. There are other aspects to consider as well. Micro-inverters typically have 25 year warranties while a standard inverters typically have 5 or 10 year warranties. The reliability of micro-inverters was in questions several years ago, but the technology now has caught up with the industry and the long warranties on the micro-inverters shows the confidence the manufacturers have in their products.

5. Micro-inverters and the add-on optimizers both offer an additional perk in system monitoring as well. With either of these devices, you have the ability to track the production of each individual panel, while with a standard inverter you only can track the production of the whole system.

If you were to expand your system in the future, micro-inverters are simple to add one at a time. However, with a standard inverter, it would be more costly to add another full unit.

To sum it all up, micro-inverters are definitely a value-add, but are only recommended if you have panels facing multiple orientations or you have shading issues. Otherwise, the less expensive standard inverter is usually more cost effective.

Disadvantages of Micro Inverters

1. The main disadvantage of micro inverters is price. They are typically a $1000 or so more expensive than a string inverter on a standard 5kW residential solar installation.

2. The second disadvantage is that you have as many inverters on your roof as you have solar panels.

String Invertor

Delta string inverter

Solar panels are installed in rows, each on a "string." For example, if you have 25 panels you may have 5 rows of 5 panels. Multiple strings are connected to one string inverter. Each string carries the DC power the solar panels produce to the string inverter where it's converted into usable AC power consumed as electricity. Depending on the size of the installation, you may have several string inverters each receiving DC power from a few strings.

String inverters have been around for a long time and are good for installations without shading issues and in which panels are positioned on a single plane so do not face different directions. If an installation uses string inverters and even one panel is shaded for a portion of the day reducing its performance, the output of every panel on the string is reduced to the struggling panels' level. Though string inverters aren't able to deal with shading issues, the technology is trusted and proven and they are less expensive than systems with micro inverters. String inverters are commonly used in residential and commercial applications. Also, as technology improves allowing string inverters to have greater power density in smaller sizes, string inverters are becoming a popular alternative over central inverters in small utility installations smaller than 1 MW.

String inverters can also be paired with power optimizers, an option that is gaining popularity. Power optimizers are module-level power electronics meaning they are installed at the module level, so each solar panel has one. Some panel manufacturers integrate their products with power optimizers and sell them as one solution known as a Smart Module. This can make installation easier. Power optimizers are able to mitigate effects of shading that string inverters alone cannot. They condition the DC electricity before sending it to the inverter, which results in a higher overall efficiency than using a string inverter alone. Power optimizers offer similar benefits as microinverters, but tend to be less expensive and so can be a good option between using strictly string inverters or microinverters.

Advantages of using String Inverters

String inverters are the most common type of solar inverter and have been around for decades. The main reason for their success is their robustness and durability, and a few years ago they weren't any alternatives on the market.

Ease of Maintenance and Troubleshooting

Most of the string inverters are installed in a common room, mounted on the wall. This facilitates the ease of maintenance, as one does not have to go back and forth to the field for troubleshooting. This is a serious benefit in places with extremely cold or hot climates.

Trusted Technology

String inverters have been around for decades. As a result of this, most string inverters are reliable and it's a technology well understood by electricians.

Lower Costs

When it comes to the costs, string inverters are currently cheaper than micro inverters. You only need one string inverter per installation, while for micro inverters , one would need one inverter per panel.

Disadvantages of String Inverters

Single Point of Failure

If the inverter breaks down, the whole solar array will be inoperable. This could be a significant loss of electricity production.

Expandability

Unless you buy an oversized string inverter, expandability in the future is restricted as once installed for a certain rating, the rating of the inverter could not be altered. In this case you would need to buy an additional inverter.

Solar Power Inverter Symbol

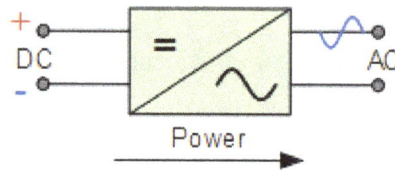

In practical terms, the inverter allows us to run electric drills, computers, vacuum cleaners, mains lighting, and most electrical appliances that can be plugged into the wall sockets. If the power inverter is big enough, then larger appliances such as freezers, refrigerators, and washing machines can also be used. All these standard 120 or 240 volt AC appliances can be powered directly from either the PV solar array, or by converting the power stored in backup batteries using the appropriately sized solar power inverter. Inverter operation is quiet and its output power is available whenever it is needed so now stand alone battery systems can run just about any standard commercial appliance, 24 hours a day.

While a single inverter may well be sufficient for a domestic installation, multiple units become the norm as we advance up the power scale and their efficiency, reliability, and safety are major concerns of the system designer.

Solar Power Inverter Configurations

Central Inverter Configuration

Several branches of the array are connected together in parallel. The complete output of the array is converted to AC through a single central solar power inverter and then fed to the grid. The single inverter is presented with a DC input voltage and current which may be quite large depending upon the configuration of the array.

This type of inverter configuration gives good efficiency, low cost, average reliability and since the PV panels within the same array are evenly matched, the maximum power point tracking (MPPT) selected by the inverter for the whole array ensures that all the PV panels operate at, or close too, their maximum power output.

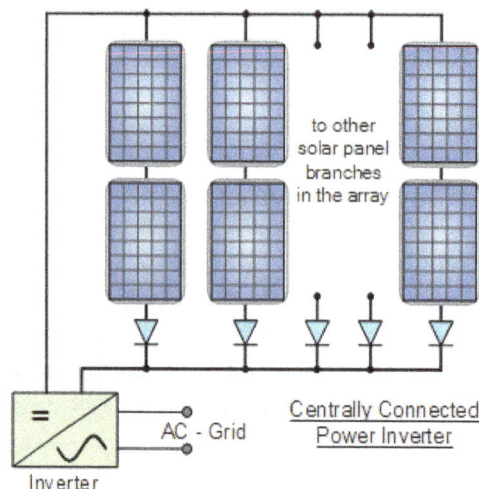

Branch Inverter Configuration

Each branch or string has its own inverter attached. Then each single branch can have a different number of PV panels, different panel types, positions, orientations or suffer from full or partial shading. The result is that each inverter produces a different power output relative to its connected array.

Therefore the array cannot be efficiently characterised by one single maximum power point (MPP), as each inverter will operate at a different maximum power point with respect to the others. The main advantage of this type of power inverter configuration is that each solar branch can be at a different location or position and not all together in one single array.

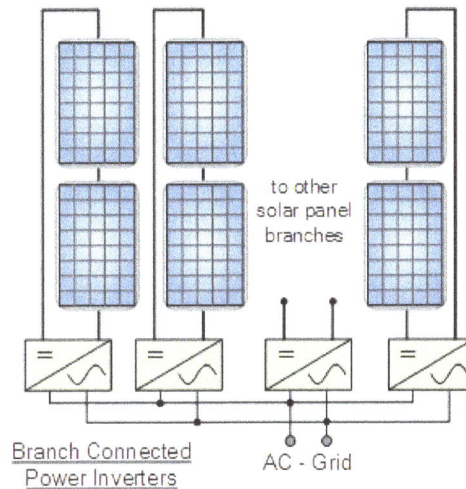

to other
solar panel
branches

Branch Connected
Power Inverters   AC - Grid

Individual Inverter Configuration

Each photovoltaic solar panel has its own power inverter. This enables the inverter to select the optimum power point for the panel giving very good efficiency but at a higher cost per kWp. More components in the array means lower reliability and more maintenance.

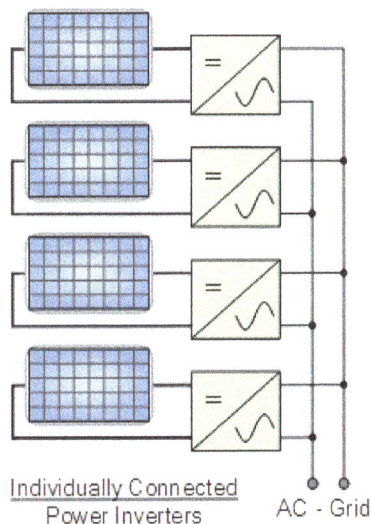

Individually Connected
Power Inverters   AC - Grid

An increasing number of solar panel manufacturers are offering individual PV panels with solar power inverters built directly into the PV panel, making each solar panel its own complete AC power source allowing it to be plugged directly into the mains grid.

Grid connected solar power inverters synchronise the electricity they produce with the local grids AC grade electricity, allowing the system to feed the solar made electricity directly into the grid, usually through a second electricity kWh "net" meter. Most grid connected power inverters are designed to operate without backup batteries, but battery based inverter models also is available. The battery based inverters for use in both stand alone and grid connected solar systems; generally include an inbuilt battery charger, which is capable of charging a battery bank directly from the grid during cloudy or bad weather.

High quality solar power inverters are available in sizes from a few 100 watts, for powering lights, laptops and games consoles from your car, up to tens of kilowatts, for powering large residential solar system with grid connected inverters being designed to automatically shut down when there is no grid power available for safety reasons. Solar inverters are available in a wide range of power sizes and voltage ratings to suit just about every combinations of installation but there are basically three kinds of DC to AC solar power inverter: square wave, modified sine wave, and pure sine wave.

Inverter Output Waveforms

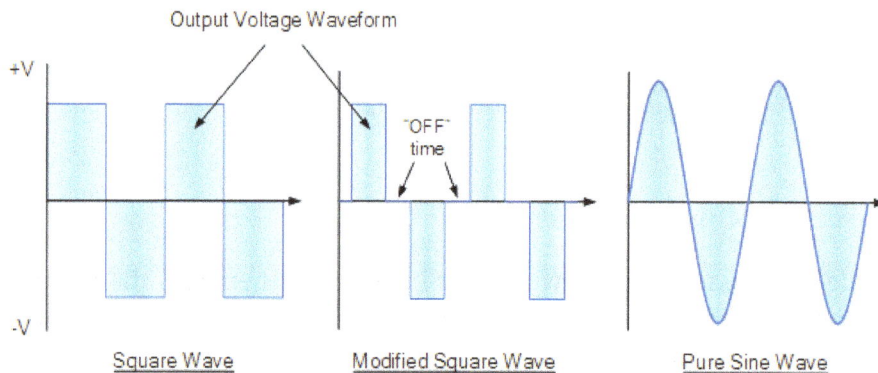

Solar Power Inverter Selection

After the solar PV panels themselves, solar power inverters are the next most important part of a grid connected PV system and therefore the DC input power rating of the inverter should be selected to match the PV panel or array. Generally, power inverters are selected for a particular solar system based on the maximum load, the maximum surge required, AC output voltage required, input battery voltage and any optional features needed. The size of an inverter is measured by its maximum continuous output in watts and this rating must be larger than the total wattage of all of the AC loads connected at the same time.

Also electrical appliances such as washing machines, dries, fridge's and freezers which use electric motors require more power to start themselves than they require to run. This high starting power consumption can be more than twice the normal power consumption so this must be considered when sizing an inverters wattage. Most power inverters are capable of delivering three to five times their rated wattage for short term surges and overload conditions.

Lets assume that we have calculated the total AC power consumption of our home and that we would need a 2,500W or 2.5kW solar inverter. The PV solar panels we are interested in are 24 volt monocrystalline silicon panels rated at 140 Watts peak. Then dividing 2,500W by 140 watts means that 18 PV panels will be needed, yielding 2,520 watts in total. But how do we connect these 18 panels to the inverter. We know from previous tutorials that PV solar panels can be connected together just like batteries, and in a series combination the voltage adds, with a constant current through each panel, and in a parallel combination the current adds with a constant voltage across each panel.

We first need to calculate how many modules can be connected together in a series branch. The datasheet for our inverter tells us that the maximum power point tracking (MPPT) input voltage is between 175 and 480 volts at a maximum current of 15 amps. The open circuit voltage (Voc) of each 24 volt PV panel at 25°C is given as 36.8 Volts. Then the maximum number of panels we can connect together in a single series branch is calculated as: 480/36.8 = 13 panels. Likewise, the minimum number of 24 volt PV panels required to keep the MPPT tracking voltage above the minimum 175 volts is there calculated as: 175/24 = 7.3, or 8 panels.

Then to keep within the inverters input voltage limits for our simple example so that the PV array voltage is not lower than 175V or greater than 480V requires an array branch of between 8 and 13 solar PV panels. Since our calculated array consists of 18 panels, two branches or strings of 9, 24 volt PV panels each is acceptable. The short circuit current, Isc of our 24 volt monocrystalline silicon panels is given as 5.8 amps. Two branches will therefore give a total maximum current of 11.6 amps, well within the inverters specification.

So to determine the number of panels that can be connected on one series PV branch, check that the sum of the open circuit voltage of all the panels does not exceed the maximum power point tracking DC input voltage and that the minimum number of panels in the series branch does not fall below the minimum MPPT voltage not forgetting that the voltage in a series branch varies up and down with temperature. Also check that the short circuit current (Isc) of the array is less than the maximum DC input current of the power inverter.

Working Principle of a Solar Inverter

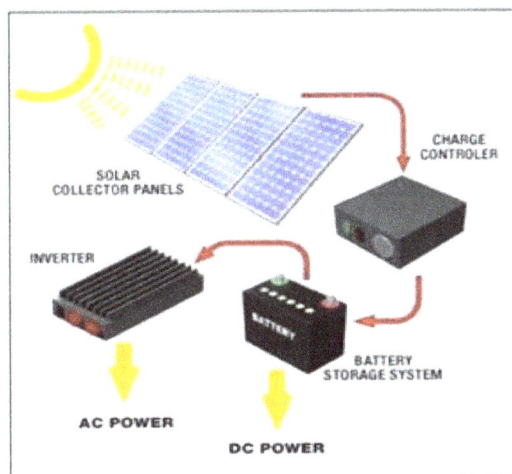

A step up transformer is used in between the devices. At this stage, the AC power output is obtained, which is either consumed by the local loads or can be fed to the grid. A few manufacturers have started manufacturing transformer-less inverters which have higher efficiencies as compared to inverters having transformers.

In any solar power inverter, a micro-controller chip is present which is programmed to execute various complex algorithms in a precise manner. The controller maximizes the power output from the photovoltaic panels through the maximum power point tracking (MPPT) algorithm.

Solar Cable

Solar cables for use in all photovoltiac systems for cabling of the solar modules and as a connection to the AC/DC inverter.

Types of Solar Cable

Wire types vary in conductor material and insulation.

Aluminum or Copper: The two common conductor materials used in residential and commercial solar installations are copper and aluminum. Copper has a greater conductivity than aluminum, thus it carries more current than aluminum at the same size.

Aluminum may be weakened during installation especially during bending; however it is less expensive than copper wires. It is not used (not permitted) for interior home wiring, as they are used in larger gauges for underground or overhead service entrances and for commercial operations.

Solid or Stranded: The cable could be solid or stranded, where stranded wires consist of many small wires that allow wire to be flexible. This type is recommended for larger sizes. The current tends to flow on the outside of the wire, thus stranded wires have slightly better conductivity as there is more wire surface.

Color

Electrical wire insulation is color coded to designate its function and use. For troubleshooting and repair, understanding the coding is essential. The wiring label differs according to AC or DC current.

Here is a simple table for color coding.

Alternating Current (AC)		Direct Current (DC)	
Color	Application	Color	Application
Black, Red or Other Color	Un-Grounded Hot	Red	Positive
White	Grounded Conductor	White	Negative or Grounded Conductor
Green or Bare	Equipment Ground	Green or Bare	Equipment Ground

Features

One common factor for most of the photovoltaic power systems is outdoor use, characterized by high temperatures and high UV radiation. Single-core cables with a maximum permissible DC voltage of 1.8 kV Umax. The phase to ground DC voltage rating must be Uo1.5kVDC and a temperature range from -40°C to +90°C ambient, 120°C on the conductor for 25 year service life against thermal ageing. Ambient temperature and conductor temperature is derived from the Arrhenius law for ageing of polymers - ageing of polymers doubles for every 10°C rise.

DC string cables must be class II double insulated to protect against short circuits and ground faults.

Small Scale Systems with String Inverters

A three-core AC cable is used for connection to the grid if a single-phase inverter is used, and a five-core cable is used for three-phase feed-in.

Large Scale System Wiring with Central Inverters

Referred to as the Main DC, larger power collector cables are used to interconnect from the Generator box also referred to as the DC combiner to the central inverter. IEC 62548 states that these cables must be shielded when over 50m in length. Central inverters are large frequency converters and unshielded cables would cause EMC issues throughout the whole solar farm which can behave like a capacitor.

Insulation

The cable's insulation must be able to withstand thermal and mechanical loads. As a consequence, plastics which have been cross-linked are increasingly used today. The insulation and jacket materials are extremely resistant to weathering, UV-radiation and abrasion . Additionally, it is salt water resistant and resistant to acids and alkaline solutions. It is suitable for fixed installation as well as for moving applications without tensile load. It is especially designed for outdoor use, which means direct sun radiation and air humidity, but due to the halogen free flame retardant cross-linked jacket material the cable can also be installed in dry and humid conditions indoors.

DC Connection

Individual modules are connected using cables to form the PV generator. The module cables are connected into a string which leads into the generator junction box, and a main DC cable connects the generator junction box to the inverter. In order to eliminate the risk of ground faults and short circuits, the positive and negative cables, each with double insulation, are laid separately.

Loss Minimization

The cross-section of the cables should be proportioned such that losses incurred in nominal operation do not exceed 1%. String cables usually have a cross-section of 4 to 10 mm^2.

Solar Tracker

Solar tracker is a system that positions an object at an angle relative to the Sun. The most-common applications for solar trackers are positioning photovoltaic (PV) panels (solar panels) so that they remain perpendicular to the Sun's rays and positioning space telescopes so that they can determine the Sun's direction. PV solar trackers adjust the direction that a solar panel is facing according to the position of the Sun in the sky. By keeping the panel perpendicular to the Sun, more sunlight strikes the solar panel, less light is reflected, and more energy is absorbed. That energy can be converted into power.

Solar panels in a field in La Calahorra, Granada, Spain

Solar tracking uses complex instruments to determine the location of the Sun relative to the object being aligned. These instruments typically include computers, which can process complicated algorithms that enable the system to track the Sun, and sensors, which provide information to a computer about the Sun's location or, when attached to a solar panel with a simple circuit board, can track the Sun without the need for a computer.

A solar tracker adjusting the direction of a solar panel, keeping the panel perpendicular to the Sun in order to maximize the amount of sunlight that strikes the panel.

Studies have shown that the angle of light affects a solar panel's power output. A solar panel that is exactly perpendicular to the Sun produces more power than a solar panel that is not perpendicular. Small angles from perpendicular have a smaller effect on power output than larger angles. In addition, Sun angle changes north to south seasonally and east to west daily. As a result, although tracking east to west is important, north to south tracking has a less-significant impact.

Solar trackers provide significant advantages for renewable energy. With solar tracking, power output can be increased by about 30 to 40 percent. The increase in power output promises to open new markets for solar power. However, solar trackers have several important disadvantages. A static solar panel may have a warranty that spans decades and may require little to no maintenance. Solar trackers, on the other hand, have much shorter warranties and require one or more actuators to move the panel. These moving parts increase installation costs and reduce reliability; active tracking systems may also use a small amount of energy (passive systems do not require additional energy). Computer-based algorithm solar trackers are more expensive, require additional maintenance, and become obsolete much faster than static solar panels, since they use fast-evolving electronic components with parts that may be difficult to replace in relatively short periods of time.

In most solar trackers, light dependent resistors (LDRs) are used as sensor. Their difference in output is used to generate error signals. In these systems, for instance, when the outputs of eastern and western LDRs become equal, the east to west tracking ends. A computer or a processor calculates the sun's position from formulae or algorithms using its time/date and geographical information to send signals to the motor orient the apparatus in such direction where illumination of sensors becomes equal and balanced.

In some high end solar trackers, a feedback system is employed wherein the output signal of various processes is sent back as the input to the system that it is controlling. Consequently, it can correct any errors and compensate for disturbances in the system.

Basic Concept

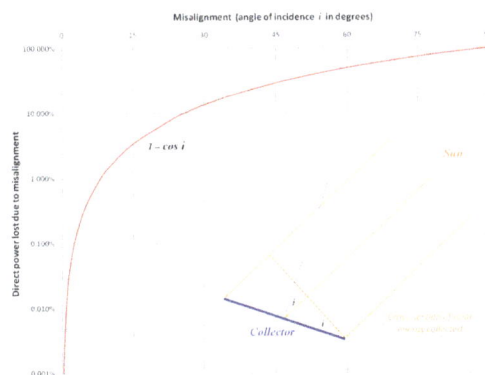

The effective collection area of a flat-panel solar collector varies with the cosine of the misalignment of the panel with the Sun.

Sunlight has two components, the "direct beam" that carries about 90% of the solar energy, and the "diffuse sunlight" that carries the remainder – the diffuse portion is the blue sky on a clear day, and is a larger proportion of the total on cloudy days. As the majority of the energy is in the direct beam, maximizing collection requires the Sun to be visible to the panels for as long as possible. However, please note that in more cloudy areas the ratio of direct vs. difuse light can be as low as 60%:40% or even lower.

The energy contributed by the direct beam drops off with the cosine of the angle between the incoming light and the panel. In addition, the reflectance (averaged across all polarizations) is approximately constant for angles of incidence up to around 50°, beyond which reflectance degrades rapidly.

Direct power lost (%) due to misalignment (angle i) where
Lost = 1 - cos(i)

i	Lost	i	hours	Lost
0°	0%	15°	1	3.4%
1°	0.015%	30°	2	13.4%
3°	0.14%	45°	3	30%
8°	1%	60°	4	>50%
23.4°	8.3%	75°	5	>75%

For example, trackers that have accuracies of ± 5°Can deliver greater than 99.6% of the energy delivered by the direct beam plus 100% of the diffuse light. As a result, high accuracy tracking is not typically used in non-concentrating PV applications.

The purpose of a tracking mechanism is to follow the Sun as it moves across the sky. In the following sections, in which each of the main factors are described in a little more detail, the complex path of the Sun is simplified by considering its daily east-west motion separately from its yearly north-south variation with the seasons of the year.

Solar Energy

The amount of solar energy available for collection from the direct beam is the amount of light intercepted by the panel. This is given by the area of the panel multiplied by the cosine of the angle of incidence of the direct beam. Or put another way, the energy intercepted is equivalent to the area of the shadow cast by the panel onto a surface perpendicular to the direct beam.

This cosine relationship is very closely related to the observation formalized in 1760 by Lambert's cosine law. This describes that the observed brightness of an object is proportional to the cosine of the angle of incidence of the light illuminating it.

Reflective losses

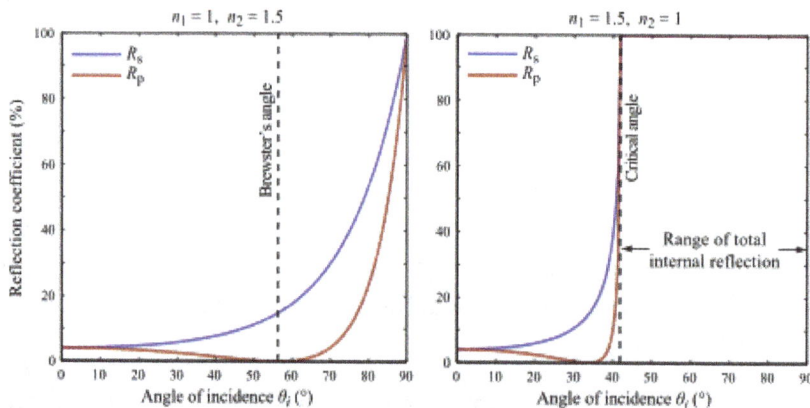

Variation of reflectance with angle of incidence

Not all of the light intercepted is transmitted into the panel - a little is reflected at its surface. The amount reflected is influenced by both the refractive index of the surface material and the angle of incidence of the incoming light. The amount reflected also differs depending on the polarization of the incoming light. Incoming sunlight is a mixture of all polarizations. Averaged over all polarizations, the reflective losses are approximately constant up to angles of incidence up to around 50° beyond which it degrades rapidly.

Daily East-west Motion of the Sun

The Sun travels through 360 degrees east to west per day, but from the perspective of any fixed location the visible portion is 180 degrees during an average 1/2 day period (more in spring and summer; less, in fall and winter). Local horizon effects reduce this somewhat, making the effective motion about 150 degrees. A solar panel in a fixed orientation between the dawn and sunset extremes will see a motion of 75 degrees to either side, and thus, according to the table above, will lose over 75% of the energy in the morning and evening. Rotating the panels to the east and west can help recapture those losses. A tracker that only attempts to compensate for the east-west movement of the Sun is known as a single-axis tracker.

Seasonal North-south Motion of the Sun

Due to the tilt of the Earth's axis, the Sun also moves through 46 degrees north and south during a year. The same set of panels set at the midpoint between the two local extremes will thus see the Sun move 23 degrees on either side. Thus according to the above table, an optimally aligned single-axis tracker will only lose 8.3% at the summer and winter seasonal extremes, or around 5% averaged over a year. Conversely a vertically or horizontally aligned single-axis tracker will lose considerably more as a result of these seasonal variations in the Sun's path. For example, a vertical tracker at a site at 60° latitude will lose up to 40% of the available energy in summer, while a horizontal tracker located at 25° latitude will lose up to 33% in winter.

A tracker that accounts for both the daily and seasonal motions is known as a dual-axis tracker. Generally speaking, the losses due to seasonal angle changes is complicated by changes in the length of the day, increasing collection in the summer in northern or southern latitudes. This biases collection toward the summer, so if the panels are tilted closer to the average summer angles, the total yearly losses are reduced compared to a system tilted at the spring/fall solstice angle (which is the same as the site's latitude).

There is considerable argument within the industry whether the small difference in yearly collection between single and dual-axis trackers makes the added complexity of a two-axis tracker worthwhile.

Other Factors

Clouds

The above models assume uniform likelihood of cloud cover at different times of day or year. In different climate zones cloud cover can vary with seasons, affecting the averaged performance figures described above. Alternatively, for example in an area where cloud cover on average builds up during the day, there can be particular benefits in collecting morning sun.

Atmosphere

The distance that sunlight has to travel through the atmosphere increases as the sun approaches the horizon, as the sunlight has to travel diagonally through the atmosphere. As the path length through the atmosphere increases, the solar intensity reaching the collector decreases. This increasing path length is referred to as the air mass (AM) or air mass coefficient, where AM0 is at the top of the atmosphere, AM1 refers to the direct vertical path down to sea-level with Sun overhead, and AM greater than 1 refers to diagonal paths as the Sun approaches the horizon.

Even though the sun may not feel particularly hot in the early mornings or during the winter months, the diagonal path through the atmosphere has a less than expected impact on the solar intensity. Even when the sun is only 15° above the horizon the solar intensity can be around 60% of its maximum value, around 50% at 10° and 25% at only 5° above the horizon. Therefore, trackers can deliver benefit by collecting the significant energy available when the Sun is close to the horizontal is this is possible.

Solar Cell Efficiency

Of course the underlying power conversion efficiency of a photovoltaic cell has a major influence on the end result, regardless of whether tracking is employed or not. Of particular relevance to the benefits of tracking are the following:

- Molecular Structure

Much research is aimed at developing surface materials to guide the maximum amount of energy down into the cell and minimize reflective losses.

- Temperature

Photovoltaic solar cell efficiency decreases with increasing temperature, at the rate of about 0.4%/°C. For example, 20% higher efficiency at 10°C in early morning or winter as compared with 60°C in the heat of the day or summer. Therefore, trackers can deliver additional benefit by collecting early morning and winter energy when the cells are operating at their highest efficiency.

Types of Solar Collector

Solar collectors may be:

- Non-concentrating flat-panels, usually photovoltaic or hot-water,
- Concentrating systems, of a variety of types.

Solar collector mounting systems may be fixed (manually aligned) or tracking. Different types of solar collector and their location (latitude) require different types of tracking mechanism. Tracking systems may be configured as:

- Fixed collector / moving mirror - i.e. *Heliostat*
- Moving collector

Non-tracking Fixed Mount

Residential and small-capacity commercial or industrial rooftop solar panels and solar water heater panels are usually fixed, often flush-mounted on an appropriately facing pitched roof. Advantages of fixed mounts over trackers include the following:

- Mechanical Advantages: Simple to manufacture, lower installation and maintenance costs.

- Wind-loading: it is easier and cheaper to provision a sturdy mount; all mounts other than fixed flush-mounted panels must be carefully designed having regard to wind loading due to greater exposure.

- Indirect light: approximately 10% of the incident solar radiation is diffuse light, available at any angle of misalignment with the Sun.

- Tolerance to misalignment: effective collection area for a flat-panel is relatively insensitive to quite high levels of misalignment with the Sun for example even a 25° misalignment reduces the direct solar energy collected by less than 10%.

Fixed mounts are usually used in conjunction with non-concentrating systems, however an important class of non-tracking concentrating collectors, of particular value in the 3rd world, are portable solar cookers. These utilize relatively low levels of concentration, typically around 2 to 8 Suns and are manually aligned.

Trackers

Even though a fixed flat-panel can be set to collect a high proportion of available noon-time energy, significant power is also available in the early mornings and late afternoons when the misalignment with a fixed panel becomes excessive to collect a reasonable proportion of the available energy. For example, even when the Sun is only 10° above the horizon the available energy can be around half the noon-time energy levels (or even greater depending on latitude, season, and atmospheric conditions).

Thus, the primary benefit of a tracking system is to collect solar energy for the longest period of the day, and with the most accurate alignment as the Sun's position shifts with the seasons.

In addition, the greater the level of concentration employed, the more important accurate tracking becomes, because the proportion of energy derived from direct radiation is higher, and the region where that concentrated energy is focused becomes smaller.

Fixed Collector/Moving Mirror

Many collectors cannot be moved, for example high-temperature collectors where the energy is recovered as hot liquid or gas (e.g. steam). Other examples include direct heating and lighting of buildings and fixed in-built solar cookers, such as Scheffler reflectors. In such cases it is necessary to employ a moving mirror so that, regardless of where the Sun is positioned in the sky, the Sun's rays are redirected onto the collector.

Due to the complicated motion of the Sun across the sky, and the level of precision required to correctly aim the Sun's rays onto the target, a heliostat mirror generally employs a dual axis tracking system, with at least one axis mechanized. In different applications, mirrors may be flat or concave.

Moving Collector

Trackers can be grouped into classes by the number and orientation of the tracker's axes. Compared to a fixed mount, a single axis tracker increases annual output by approximately 30%, and a dual axis tracker an additional 10-20%.

Photovoltaic trackers can be classified into two types: standard photovoltaic (PV) trackers and concentrated photovoltaic (CPV) trackers. Each of these tracker types can be further categorized by the number and orientation of their axes, their actuation architecture and drive type, their intended applications, their vertical supports and foundation.

Floating Ground Mount

Solar trackers can be built using a "floating" foundation, which sits on the ground without the need for invasive concrete foundations. Instead of placing the tracker on concrete foundations, the tracker is placed on a gravel pan that can be filled with a variety of materials, such as sand or gravel, to secure the tracker to the ground. These "floating" trackers can sustain the same wind load as a traditional fixed mounted tracker. The use of floating trackers increases the number of potential sites for commercial solar projects since they can be placed on top of capped landfills or in areas where excavated foundations are not feasible.

Non-concentrating Photovoltaic (PV) Trackers

Photovoltaic panels accept both direct and diffuse light from the sky. The panels on standard photovoltaic trackers gather both the available direct and diffuse light. The tracking functionality in standard photovoltaic trackers is used to minimize the angle of incidence between incoming light and the photovoltaic panel. This increases the amount of energy gathered from the direct component of the incoming sunlight.

The physics behind standard photovoltaic (PV) trackers works with all standard photovoltaic module technologies. These include all types of crystalline silicon panels (either mono-Si, or multi-Si) and all types of thin film panels (amorphous silicon, CdTe, CIGS, microcrystalline).

Concentrator Photovoltaic (CPV) Trackers

The optics in CPV modules accept the direct component of the incoming light and therefore must be oriented appropriately to maximize the energy collected. In low concentration applications a portion of the diffuse light from the sky can also be captured. The tracking functionality in CPV modules is used to orient the optics such that the incoming light is focused to a photovoltaic collector.

3-megawatt CPV plant using dual axis trackers in Golmud, China

200-kilowatt CPV modules on dual axis tracker in Qingdao, China

CPV modules that concentrate in one dimension must be tracked normal to the Sun in one axis. CPV modules that concentrate in two dimensions must be tracked normal to the Sun in two axes.

Accuracy Requirements

The physics behind CPV optics requires that tracking accuracy increase as the systems concentration ratio increases. However, for a given concentration, nonimaging optics provide the widest possible acceptance angles, which may be used to reduce tracking accuracy.

In typical high concentration systems tracking accuracy must be in the ± 0.1° range to deliver approximately 90% of the rated power output. In low concentration systems, tracking accuracy must be in the ± 2.0° range to deliver 90% of the rated power output. As a result, high accuracy tracking systems are typical.

Technologies Supported

Concentrated photovoltaic trackers are used with refractive and reflective based concentrator systems. There are a range of emerging photovoltaic cell technologies used in these systems. These range from conventional, crystalline silicon-based photovoltaic receivers to germanium-based triple junction receivers.

Single Axis Trackers

Single axis trackers have one degree of freedom that acts as an axis of rotation. The axis of rotation of single axis trackers is typically aligned along a true North meridian. It is possible to align them in any cardinal direction with advanced tracking algorithms. There are several common implementations of single axis trackers. These include horizontal single axis trackers (HSAT), horizontal single axis tracker with tilted modules (HTSAT), vertical single axis trackers (VSAT), tilted single axis trackers (TSAT) and polar aligned single axis trackers (PSAT). The orientation of the module with respect to the tracker axis is important when modeling performance.

Horizontal

Horizontal Single Axis Tracker (HSAT)

4MW horizontal single axis tracker in Vellakoil, Tamil Nadu, India

Horizontal Single Axis tracker with Tilted Modules in Xitieshan, China. Commissioned in July 2014.

The axis of rotation for horizontal single axis tracker is horizontal with respect to the ground. The posts at either end of the axis of rotation of a horizontal single axis tracker can be shared between trackers to lower the installation cost. This type of solar tracker is most appropriate for low latitude regions. Field layouts with horizontal single axis trackers are very flexible. The simple geometry means that keeping all of the axes of rotation parallel to one another is all that is required for appropriately positioning the trackers with respect to one another. Appropriate spacing can maximize the ratio of energy production to cost, this being dependent upon local terrain and shading conditions and the time-of-day value of the energy produced. Backtracking is one means of computing the disposition of panels. Horizontal trackers typically have the face of the module oriented parallel to the axis of rotation. As a module tracks, it sweeps a cylinder that is rotationally symmetric around the axis of rotation. In single axis horizontal trackers, a long horizontal tube is supported on bearings mounted upon pylons or frames. The axis of the tube is on a north–south line. Panels are mounted upon the tube, and the tube will rotate on its axis to track the apparent motion of the Sun through the day.

Horizontal Single Axis Tracker with Tilted Modules (HTSAT)

In HSAT, the modules are mounted flat at 0 degrees, while in HTSAT, the modules are installed at a certain tilt. It works on same principle as HSAT, keeping the axis of tube horizontal in north–south line and rotates the solar modules east to west throughout the day. These trackers are usually suitable in high latitude locations but do not take as much land space as consumed by Vertical single axis tracker (VSAT). Therefore, it brings the advantages of VSAT in a horizontal tracker and minimizes the overall cost of solar project.

Vertical

Vertical Single Axis Tracker (VSAT)

The axis of rotation for vertical single axis trackers is vertical with respect to the ground. These trackers rotate from East to West over the course of the day. Such trackers are more effective at high latitudes than are horizontal axis trackers. Field layouts must consider shading to avoid unnecessary energy losses and to optimize land utilization. Also optimization for dense packing is limited due to the nature of the shading over the course of a year. Vertical single axis trackers typically have the face of the module oriented at an angle with respect to the axis of rotation. As a module tracks, it sweeps a cone that is rotationally symmetric around the axis of rotation.

Tilted

Tilted Single Axis Tracker (TSAT)

Tilted single axis tracker in Siziwangqi, China.

All trackers with axes of rotation between horizontal and vertical are considered tilted single axis trackers. Tracker tilt angles are often limited to reduce the wind profile and decrease the elevated end height. With backtracking, they can be packed without shading perpendicular to their axis of rotation at any density. However, the packing parallel to their axes of rotation is limited by the tilt angle and the latitude. Tilted single axis trackers typically have the face of the module oriented parallel to the axis of rotation. As a module tracks, it sweeps a cylinder that is rotationally symmetric around the axis of rotation.

Dual Axis Trackers

Dual axis trackers have two degrees of freedom that act as axes of rotation. These axes are typically normal to one another. The axis that is fixed with respect to the ground can be considered a primary axis. The axis that is referenced to the primary axis can be considered a secondary axis. There are several common implementations of dual axis trackers. They are classified by the orientation of their primary axes with respect to the ground. Two common implementations are tip-tilt dual axis trackers (TTDAT) and azimuth-altitude dual axis trackers (AADAT). The orientation of the module with respect to the tracker axis is important when modeling performance. Dual axis trackers typically have modules oriented parallel to the secondary axis of rotation. Dual axis trackers allow for optimum solar energy levels due to their ability to follow the Sun vertically and horizontally. No matter where the Sun is in the sky, dual axis trackers are able to angle themselves to be in direct contact with the Sun.

Tip–tilt

Dual axis tracker mounted on a pole

A tip–tilt dual axis tracker (TTDAT) is so-named because the panel array is mounted on the top of a pole. Normally the east–west movement is driven by rotating the array around the top of the pole. On top of the rotating bearing is a T- or H-shaped mechanism that provides vertical rotation of the panels and provides the main mounting points for the array. The posts at either end of the primary axis of rotation of a tip–tilt dual axis tracker can be shared between trackers to lower installation costs.

Other such TTDAT trackers have a horizontal primary axis and a dependent orthogonal axis. The vertical azimuthal axis is fixed. This allows for great flexibility of the payload connection to the ground mounted equipment because there is no twisting of the cabling around the pole.

Field layouts with tip–tilt dual axis trackers are very flexible. The simple geometry means that keeping the axes of rotation parallel to one another is all that is required for appropriately positioning the trackers with respect to one another. Normally the trackers would have to be positioned at fairly low density in order to avoid one tracker casting a shadow on others when the Sun is low in the sky. Tip-tilt trackers can make up for this by tilting closer to horizontal to minimize up-Sun shading and therefore maximize the total power being collected.

The axes of rotation of many tip–tilt dual axis trackers are typically aligned either along a true north meridian or an east–west line of latitude.

Given the unique capabilities of the tip-tilt configuration and the appropriated controller totally automatic tracking is possible for use on portable platforms. The orientation of the tracker is of no importance and can be placed as needed.

Azimuth-altitude dual axis tracker - 2 axis solar tracker, Toledo, Spain.

Azimuth-altitude

An azimuth–altitude (or alt-azimuth) dual axis tracker (AADAT) has its primary axis (the azimuth axis) vertical to the ground. The secondary axis, often called elevation axis, is then typically normal to the primary axis. They are similar to tip-tilt systems in operation, but they differ in the way the array is rotated for daily tracking. Instead of rotating the array around the top of the pole, AADAT systems can use a large ring mounted on the ground with the array mounted on a series of rollers. The main advantage of this arrangement is the weight of the array is distributed over a portion of the ring, as opposed to the single loading point of the pole in the TTDAT. This allows AADAT to support much larger arrays. Unlike the TTDAT, however, the AADAT system cannot be placed closer together than the diameter of the ring, which may reduce the system density, especially considering inter-tracker shading.

Construction and (Self-)Build

The economical balance between cost of panel and tracker is not trivial. The steep drop in cost for solar panels in the early 2010s made it more challenging to find a sensible solution. As can be seen in the attached media files, most constructions use industrial and/or heavy materials unsuitable for small or craft workshops. Even commercial offers like "Complete-Kit-1KW-Single-Axis-Solar-Panel-Tracking-System-Linear-Actuator-Electric-Controller-For-Sunlight-Solar/1279440_2037007138" have rather unsuitable solutions (a big rock) for stabilization. For a small (amateur/enthusiast) construction following criteria have to be met: economy, stability of endproduct against elemental hazards, ease of handling materials and joinery.

Tracker Type Selection

The selection of tracker type is on many factors including installation size, electric rates, government incentives, land constraints, latitude, and local weather.

Horizontal single axis trackers are typically used for large distributed generation projects and utility scale projects. The combination of energy improvement and lower product cost and lower installation complexity results in compelling economics in large deployments. In addition the strong afternoon performance is particularly desirable for large grid-tied photovoltaic systems so that production will match the peak demand time. Horizontal single axis trackers also add a substantial amount of productivity during the spring and summer seasons when the Sun is high in the sky. The inherent robustness of their supporting structure and the simplicity of the mechanism also result in high reliability which keeps maintenance costs low. Since the panels are horizontal, they can be compactly placed on the axle tube without danger of self-shading and are also readily accessible for cleaning.

A vertical axis tracker pivots only about a vertical axle, with the panels either vertical, at a fixed, adjustable, or tracked elevation angle. Such trackers with fixed or (seasonally) adjustable angles are suitable for high latitudes, where the apparent solar path is not especially high, but which leads to long days in summer, with the Sun traveling through a long arc.

Dual axis trackers are typically used in smaller residential installations and locations with very high government feed in tariffs.

Multi-mirror Concentrating PV

Reflective mirror concentrator units

This device uses multiple mirrors in a horizontal plane to reflect sunlight upward to a high temperature photovoltaic or other system requiring concentrated solar power. Structural problems and expense are greatly reduced since the mirrors are not significantly exposed to wind loads. Through the employment of a patented mechanism, only two drive systems are required for each device. Because of the configuration of the device it is especially suited for use on flat roofs and at lower latitudes. The units illustrated each produce approximately 200 peak DC watts.

A multiple mirror reflective system combined with a central power tower is employed at the Sierra SunTower, located in Lancaster, California. This generation plant operated by eSolar is scheduled to begin operations on August 5, 2009. This system, which uses multiple heliostats in a north–south alignment, uses pre-fabricated parts and construction as a way of decreasing startup and operating costs.

Drive Types

Active Tracker

Active trackers use motors and gear trains to perform solar tracking. They can use microprocessors and sensors, date and time-based algorithms, or a combination of both to detect the position of the sun. In order to control and manage the movement of these massive structures special slewing drives are designed and rigorously tested. The technologies used to direct the tracker are constantly evolving and recent developments at Google and Eternity have included the use of wire-ropes and winches to replace some of the more costly and more fragile components.

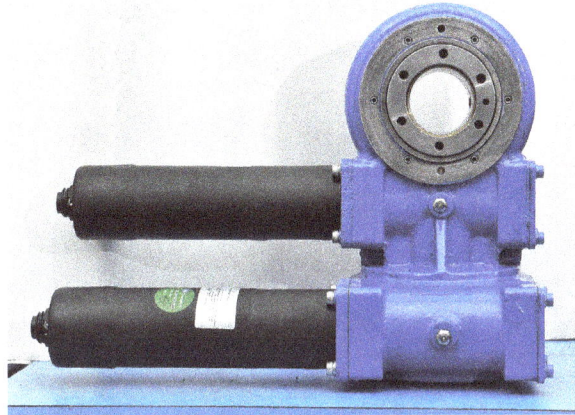

A slewing drive gearbox

Counter rotating slewing drives sandwiching a fixed angle support can be applied to create a "multi-axis" tracking method which eliminates rotation relative to longitudinal alignment. This method if placed on a column or pillar will generate more electricity than fixed PV and its PV array will never rotate into a parking lot drive lane. It will also allow for maximum solar generation in virtually any parking lot lane/row orientation, including circular or curvilinear.

Active two-axis trackers are also used to orient heliostats - movable mirrors that reflect sunlight toward the absorber of a central power station. As each mirror in a large field will have an individual orientation these are controlled programmatically through a central computer system, which also allows the system to be shut down when necessary.

Light-sensing trackers typically have two or more photosensors, such as photodiodes, configured differentially so that they output a null when receiving the same light flux. Mechanically, they should be omnidirectional (i.e. flat) and are aimed 90 degrees apart. This will cause the steepest part of their cosine transfer functions to balance at the steepest part, which translates into maximum sensitivity.

Since the motors consume energy, one wants to use them only as necessary. So instead of a continuous motion, the heliostat is moved in discrete steps. Also, if the light is below some threshold there would not be enough power generated to warrant reorientation. This is also true when there is not enough difference in light level from one direction to another, such as when clouds are passing overhead. Consideration must be made to keep the tracker from wasting energy during cloudy periods.

Passive Tracker

The most common passive trackers use a low boiling point compressed gas fluid that is driven to one side or the other (by solar heat creating gas pressure) to cause the tracker to move in response to an imbalance. As this is a non-precision orientation it is unsuitable for certain types of concentrating photovoltaic collectors but works fine for common PV panel types. These will have viscous dampers to prevent excessive motion in response to wind gusts. Shader/reflectors are used to reflect early morning sunlight to "wake up" the panel and tilt it toward the Sun, which can take nearly an hour. The time to do this can be greatly reduced by adding a self-releasing tie down that positions the panel slightly past the zenith (so that the fluid does not have to overcome gravity) and using the tie down in the evening. (A slack-pulling spring will prevent release in windy overnight conditions.)

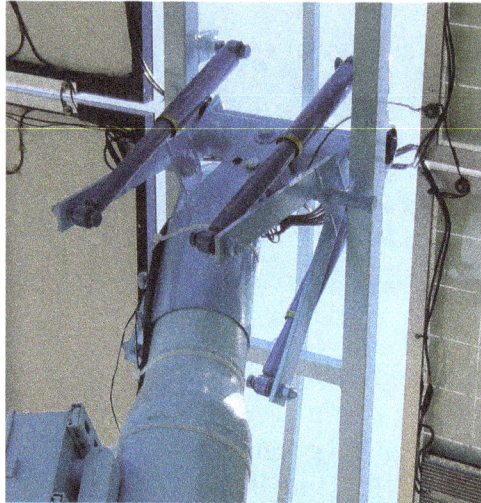

Passive tracker head in spring/summer tilt position with panels on light blue rack pivoted to morning position against stop; dark blue objects are hydraulic dampers.

A newly emerging type of passive tracker for photovoltaic solar panels uses a hologram behind stripes of photovoltaic cells so that sunlight passes through the transparent part of the module and reflects on the hologram. This allows sunlight to hit the cell from behind, thereby increasing the module's efficiency. Also, the panel does not have to move since the hologram always reflects sunlight from the correct angle towards the cells.

Manual Tracking

In some developing nations, drives have been replaced by operators who adjust the trackers. This has the benefits of robustness, having staff available for maintenance and creating employment for the population in the vicinity of the site.

Rotating Buildings

In Freiburg im Breisgau, Germany, Rolf Disch built the Heliotrop in 1996, a residential building that is rotating with the sun and has an additional dual axis photovoltaic sail on the roof. It's producing four times the amount of energy the building consumes.

The Gemini House is a unique example of a vertical axis tracker. This cylindrical house in Austria (latitude above 45 degrees north) rotates in its entirety to track the Sun, with vertical solar panels mounted on one side of the building, rotating independently, allowing control of the natural heating from the Sun.

ReVolt House is a rotating, floating house designed by TU Delft students for the Solar Decathlon Europe competition in Madrid. The house would be realized in September 2012. A closed façade turns itself towards the Sun in summer to prevent the interior space from direct heat gains. In winter, the glass façade faces the Sun to get direct sunlight in the house.

Gemini house rotates in its entirety.

Disadvantages

Trackers add cost and maintenance to the system - if they add 25% to the cost, and improve the output by 25%, the same performance can be obtained by making the system 25% larger, eliminating the additional maintenance. Tracking was very cost effective in the past when photovoltaic modules were expensive compared to today. Because they were expensive, it was important to use tracking to minimize the number of panels used in a system with a given power output. But as panels get cheaper, the cost effectiveness of tracking vs using a greater number of panels decreases.

Tracking is also not suitable for typical residential rooftop photovoltaic installations. Since tracking requires that panels tilt or otherwise move, provisions must be made to allow this. This requires that panels be offset a significant distance from the roof, which requires expensive racking and increases wind load. Also, such a setup would not make for a very aesthetically pleasing install on residential rooftops. Because of this (and the high cost of such a system), tracking is not used on residential rooftop installations, and is unlikely to ever be used in such installations. This is especially true as the cost of photovoltaic modules continues to decrease, which makes increasing the number of modules for more power the more cost-effective option. Tracking can (and sometimes is) used for residential ground mount installations, where greater freedom of movement is possible.

Tracking can also cause shading problems. As the panels move during the course of the day, it is possible that, if the panels are located too close to one another, they may shade one another due to profile angle effects. As an example, if you have several panels in a row from east to west, there will be no shading during solar noon. But in the afternoon, panels could be shaded by their west neighboring panel if they are sufficiently close. This means that panels must be spaced sufficiently far to prevent shading in systems with tracking, which can reduce the available power from a given area during the peak Sun hours. This is not a big problem if there is sufficient land area to widely space the panels. But it will reduce output during certain hours of the day (i.e. around solar noon) compared to a fixed array.

Further, single-axis tracking systems are prone to become unstable already at relatively modest wind speeds (galloping). This is due to torsional instability of single-axis solar tracking systems. Anti-galloping measures such as automatic stowing and external dampers must be implemented.

Maximum Power Point Tracking

Maximum power point tracking, frequently referred to as MPPT, is an electronic system that operates the Photovoltaic (PV) modules in a manner that allows the modules to produce all the power they are capable of. MPPT is not a mechanical tracking system that "physically moves" the modules to make them point more directly at the sun. MPPT is a fully electronic system that varies the electrical operating point of the modules so that the modules are able to deliver maximum available power. Additional power harvested from the modules is then made available as increased battery charge current. MPPT can be used in conjunction with a mechanical tracking system, but the two systems are completely different.

To understand how MPPT works, let's first consider the operation of a conventional (non-MPPT) charge controller. When a conventional controller is charging a discharged battery, it simply connects the modules directly to the battery. This forces the modules to operate at battery voltage, typically not the ideal operating voltage at which the modules are able to produce their maximum available power. The PV Module Power/Voltage/Current graph shows the traditional Current/Voltage curve for a typical 75W module at standard test conditions of 25°C cell temperature and 1000W/m² of insolation. This graph also shows PV module power delivered vs module voltage. For the example shown, the conventional controller simply connects the module to the battery and therefore forces the module to operate at 12V. By forcing the 75W module to operate at 12V the conventional controller artificially limits power production to $\approx 53\text{W}$.

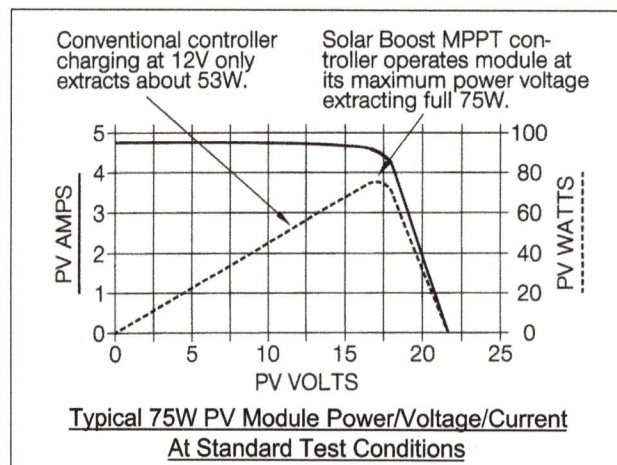

Conventional controller charging at 12V only extracts about 53W.

Solar Boost MPPT controller operates module at its maximum power voltage extracting full 75W.

Typical 75W PV Module Power/Voltage/Current
At Standard Test Conditions

Rather than simply connecting the module to the battery, the patented MPPT system in a Solar Boost charge controller calculates the voltage at which the module is able to produce maximum power. In this example the maximum power voltage of the module (V_{MP}) is 17V. The MPPT system then operates the modules at 17V to extract the full 75W, regardless of present battery voltage. A high efficiency DC-to-DC power converter converts the 17V module voltage at the controller input to battery voltage at the output. If the whole system wiring and all was 100% efficient, battery charge current in this example would be $V_{MODULE} \div V_{BATTERY} \times I_{MODULE}$, or $17V \div 12V \times 4.45A = 6.30A$. A charge current increase of 1.85A or 42% would be achieved by harvesting module power that would have been left behind by a conventional controller and turning it into useable charge current. But, nothing is 100% efficient and actual charge current increase will be somewhat lower as some power is lost in wiring, fuses, circuit breakers, and in the Solar Boost charge controller.

Actual charge current increase varies with operating conditions. As shown above, the greater the difference between PV module maximum power voltage V_{MP} and battery voltage, the greater the charge current increase will be. Cooler PV module cell temperatures tend to produce higher V_{MP} and therefore greater charge current increase. This is because V_{MP} and available power increase as module cell temperature decreases as shown in the PV Module Temperature Performance graph. Modules with a 25°C V_{MP} rating higher than 17V will also tend to produce more charge current increase because the difference between actual V_{MP} and battery voltage will be greater. A highly discharged battery will also increase charge current since battery voltage is lower, and output to the battery during MPPT could be thought of as being "constant power".

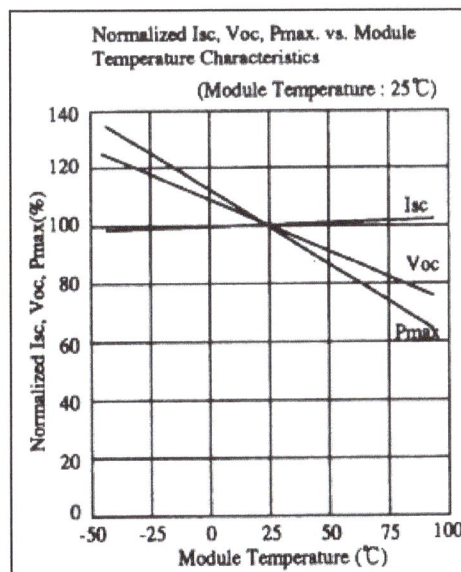

Typical PV Module Temperature Performance

Implementation

When a load is directly connected to the solar panel, the operating point of the panel will rarely be at peak power. The impedance seen by the panel derives the operating point of the solar panel. Thus by varying the impedance seen by the panel, the operating point can be moved towards peak power point. Since panels are DC devices, DC-DC converters must be utilized to transform the impedance of one circuit (source) to the other circuit (load). Changing the duty ratio of the DC-DC converter results in an impedance change as seen by the panel. At a particular impedance (or duty ratio) the operating point will be at the peak power transfer point. The I-V curve of the panel can vary considerably with variation in atmospheric conditions such as radiance and temperature. Therefore, it is not feasible to fix the duty ratio with such dynamically changing operating conditions.

MPPT implementations utilize algorithms that frequently sample panel voltages and currents, then adjust the duty ratio as needed. Microcontrollers are employed to implement the algorithms. Modern implementations often utilize larger computers for analytics and load forecasting.

Classification

Controllers can follow several strategies to optimize the power output of an array. Maximum power point trackers may implement different algorithms and switch between them based on the operating conditions of the array.

Perturb and Observe

In this method the controller adjusts the voltage by a small amount from the array and measures power; if the power increases, further adjustments in that direction are tried until power no longer increases. This is called the perturb and observe method and is most common, although this method can result in oscillations of power output. It is referred to as a *hill climbing* method, because it depends on the rise of the curve of power against voltage below the maximum power point, and the fall above that point. Perturb and observe is the most commonly used MPPT method due to its ease of implementation. Perturb and observe method may result in top-level efficiency, provided that a proper predictive and adaptive hill climbing strategy is adopted.

Incremental Conductance

In the incremental conductance method, the controller measures incremental changes in PV array current and voltage to predict the effect of a voltage change. This method requires more computation in the controller, but can track changing conditions more rapidly than the perturb and observe method (P&O). Like the P&O algorithm, it can produce oscillations in power output. This method utilizes the incremental conductance (dI/dV) of the photovoltaic array to compute the sign of the change in power with respect to voltage (dP/dV).

The incremental conductance method computes the maximum power point by comparison of the incremental conductance (I_Δ / V_Δ) to the array conductance (I / V). When these two are the same ($I / V = I_\Delta / V_\Delta$), the output voltage is the MPP voltage. The controller maintains this voltage until the irradiation changes and the process is repeated.

The incremental conductance method is based on the observation that at the maximum power point $dP/dV = 0$, and that $P = IV$. The current from the array can be expressed as a function of the voltage: $P = I(V)V$. Therefore, $dP/dV = VdI/dV + I(V)$. Setting this equal to zero yields: $dI/dV = -I(V)/V$. Therefore, the maximum power point is achieved when the incremental conductance is equal to the negative of the instantaneous conductance.

Current Sweep

The current sweep method uses a sweep waveform for the PV array current such that the I-V characteristic of the PV array is obtained and updated at fixed time intervals. The maximum power point voltage can then be computed from the characteristic curve at the same intervals.

Constant Voltage

The term "constant voltage" in MPP tracking is used to describe different techniques by different authors, one in which the output voltage is regulated to a constant value under all conditions and one in which the output voltage is regulated based on a constant ratio to the measured open circuit voltage (V_{OC}). The latter technique is referred to in contrast as the "open voltage" method by some authors. If the output voltage is held constant, there is no attempt to track the maximum power point, so it is not a maximum power point tracking technique in a strict sense, though it does have some advantages in cases when the MPP tracking tends to fail, and thus it is sometimes used to supplement an MPPT method in those cases.

In the "constant voltage" MPPT method (also known as the "open voltage method"), the power delivered to the load is momentarily interrupted and the open-circuit voltage with zero current is measured. The controller then resumes operation with the voltage controlled at a fixed ratio, such as 0.76, of the open-circuit voltage V_{OC}. This is usually a value which has been determined to be the maximum power point, either empirically or based on modelling, for expected operating conditions. The operating point of the PV array is thus kept near the MPP by regulating the array voltage and matching it to the fixed reference voltage $V_{ref}=kV_{OC}$. The value of V_{ref} may be also chosen to give optimal performance relative to other factors as well as the MPP, but the central idea in this technique is that V_{ref} is determined as a ratio to V_{OC}.

One of the inherent approximations to the "constant voltage" ratio method is that the ratio of the MPP voltage to V_{OC} is only approximately constant, so it leaves room for further possible optimization.

Comparison of Methods

Both perturb and observe, and incremental conductance, are examples of "hill climbing" methods that can find the local maximum of the power curve for the operating condition of the PV array, and so provide a true maximum power point.

The perturb and observe method requires oscillating power output around the maximum power point even under steady state irradiance.

The incremental conductance method has the advantage over the perturb and observe (P&O) method that it can determine the maximum power point without oscillating around this value. It can perform maximum power point tracking under rapidly varying irradiation conditions with higher accuracy than the perturb and observe method. However, the incremental conductance method can produce oscillations (unintentionally) and can perform erratically under rapidly changing atmospheric conditions. The sampling frequency is decreased due to the higher complexity of the algorithm compared to the P&O method.

In the constant voltage ratio (or "open voltage") method, the current from the photovoltaic array must be set to zero momentarily to measure the open circuit voltage and then afterwards set to a predetermined percentage of the measured voltage, usually around 76%. Energy may be wasted during the time the current is set to zero. The approximation of 76% as the MPP/V_{OC} ratio is not necessarily accurate. Although simple and low-cost to implement, the interruptions reduce array efficiency and do not ensure finding the actual maximum power point. However, efficiencies of some systems may reach above 95%.

MPPT Placement

Traditional solar inverters perform MPPT for the entire PV array as a whole. In such systems the same current, dictated by the inverter, flows through all modules in the string (series). Because different modules have different I-V curves and different MPPs (due to manufacturing tolerance, partial shading, etc.) this architecture means some modules will be performing below their MPP, resulting in lower efficiency.

Some companies are now placing maximum power point tracker into individual modules, allowing each to operate at peak efficiency despite uneven shading, soiling or electrical mismatch.

Data suggests having one inverter with one MPPT for a project that has east and west-facing modules presents no disadvantages when compared to having two inverters or one inverter with more than one MPPT.

Operation with Batteries

At night, an off-grid PV system may use batteries to supply loads. Although the fully charged battery pack voltage may be close to the PV panel's maximum power point voltage, this is unlikely to be true at sunrise when the battery has been partially discharged. Charging may begin at a voltage considerably below the PV panel maximum power point voltage, and an MPPT can resolve this mismatch.

When the batteries in an off-grid system are fully charged and PV production exceeds local loads, an MPPT can no longer operate the panel at its maximum power point as the excess power has no load to absorb it. The MPPT must then shift the PV panel operating point away from the peak power point until production exactly matches demand. An alternative approach commonly used in spacecraft is to divert surplus PV power into a resistive load, allowing the panel to operate continuously at its peak power point.

In a grid connected photovoltaic system, all delivered power from solar modules will be sent to the grid. Therefore, the MPPT in a grid connected PV system will always attempt to operate the PV modules at its maximum power point.

Photovoltaic Mounting System

Photovoltaic mounting system is a mounting racks to fix solar panels to the roof or the ground.

Roof-Mount Racking Systems

The most common installation of residential solar panels is a roof-mount racking system. Within this category, you'll find a few different types.

Rail-Mounting System

A rail-mounting rack system gets used on a roof sloped at an angle. You secure the rails to the roof using a bolt or screw with flashing installed either around or over the hole made by the bolt or screw to create a watertight seal preventing any leaks. Each solar panel is then attached to two rails, one on the top and one on the bottom.

Pros of rail-mounting system:

- Solar panels are secure.
- This mounting system works well in areas with high winds.

Cons of rail-mounting system:

- The pieces weigh more than a shared rail system and a rail-less system.
- The system requires penetrations into the roof.
- The arrangement is rigid for where you can install the panels.

Shared-Rail System

Similar to a rail-mounting system, the shared-rail system uses rails. However, instead of two rows of rails for each row of solar panels, you only need three rows of rails for every two rows of panels. The two rows share the middle rail, which means fewer bolts or screws get installed into the roof.

Pros of shared-rail system:

- The installation is faster and easier than a rail-mounting system.
- You can install the shared rails vertically or horizontally.
- This installation is compatible with both tile and composite roofs.

- The item uses fewer components than other rail systems and fewer penetrations than a rail-less system.

Cons of shared-rail system:

- This system weighs more than a rail-less system.

- You'll spend more on shipping costs than a rail-less system.

- Panel installation is not flexible.

Rail-less Mounting System

Another option on a sloped roof is a rail-less system. This system doesn't require the solar panels to be attached to rails. Instead, the panels themselves get attached to the roof with bolts or screws.

Pros of rail-less mounting system:

- You have more flexibility over how the solar panels get installed, including the direction.

- This system has lower costs because you don't need to have the solar panels attached to the rails for shipping.

- The arrangement overall has fewer components, which speeds up installation time, making it faster than a rail system.

Cons of rail-less mounting systems:

- You have only a small part of the panel frame attached to the roof, which, over time, can cause damage from micro fractures, resulting in a loss of output.

- This system has more penetrations than other racking systems.

Flush-Mount Systems

With this mounting system, solar panels are more flush with the roof. The design protects the roof as well as the panels, providing 2 to 4 inches between the roof and the bottom of the panel. Here, you'll get air flow to the underside of the panel while reducing wind load.

Pros of flush-mount systems:

- You usually don't need penetrations.

- This system reduces the dead load on your roof.

- The arrangement works with any type of solar panel system.

- The materials are suitable for areas that experience high winds.

Cons of flush-mount systems:

- Depending on the type of roof, you may need to make roof penetrations.

Ballasted-Mount Systems

Flat-roof mounting systems typically use this type of mount to keep the solar panels in place. However, you can use them on other roof types as well. You use weights to hold the solar panels in place.

Pros of ballasted-mount systems:

- You don't need penetrations.

- The system costs less to install, and the overall installation is faster.

Cons of ballasted-mount systems:

- This system increases the load on your roof.

- You couldn't use this system in areas with high winds.

- You'll have more difficulties getting the weights up to your roof.

- Costs may be higher to ship the system because of the extra weight.

Ground-Mount Solar-Racking Systems

If you have a larger plot of land, you may be able to have a ground-mounting system instead of a roof-mounting system. Ground-racking systems typically cost the same as roof systems on a per-watt basis and can save you money in the long run if your rooftop isn't ideal for solar panels, especially if your roof doesn't face south or is not at the right angle. You'll gain more flexibility over placing ground-mount systems, and you can adjust them to meet the energy consumption in your home.

Standard Ground Mounts

This type of solar-racking system uses metal frames driven into the ground. These frames hold the panels at a fixed angle, although some systems are manually adjusted up to three times a year so that you can shift them as the seasons and sun routes change.

Pros of standard ground mounts:

- These mounts don't require penetrations into your roof.

- The mounts are lightweight and easy to move.

- Ground mounts are generally more productive than roof mounts because they're set at the perfect angle.

- You can easily access ground mounts for cleaning and maintenance, such as removing snow in the winter.

Cons of standard ground mounts:

- You need open land to place the solar panels.

- You also must have a spot on the land where the panels can face the right angle with no interference from trees or other obstructions.

Pole Mounts

With pole-mounted racking system, multiple solar panels get attached on a single pole. These poles become raised higher off of the ground and often use a tracking system to tilt the panels automatically to make sure the panels absorb the most sunshine possible.

Pros of standard pole mounts:

- You don't need to install these types of mounts on your roof.

- These mounts are more productive than most other systems because they use a tracking system and, since they're higher up, they are able to avoid obstructions.

Cons of standard pole mounts:

- You need a sizeable area in which to place the poles away from obstructions.

- You need to install the poles so that they're more secure and can't be moved easily.

Thermophotovoltaic Generator

A thermophotovoltaic generator (TPV) is an innovative system able to convert the radiant energy of combustion into electrical energy. This conversion is realized by using photovoltaic cells. A scheme of a TPV is presented in below figure, in which the main components and energy flows are highlighted.

A TPV generator consists of a heat source, an emitter (EM), a filter (F) and an array of photovoltaic cells (PV); the combustion air pre-heating system (HX-A) which uses the combustion products is also sketched in figure. The thermal production of the TPV is realized by the heat exchangers HX-PV and HX-CP, which respectively recover the heat from the cooling of PV cells and the exhaust combustion products.

The main advantages of this energy system can be found in the:

i. High fuel utilization factor (close to the unity thanks to the recovery of the most of the thermal losses, making it possible to use the TPV system as a combined heat and power system),

ii. Low produced noise levels (due to the absence of moving parts),

iii. Easy maintenance (similar to a common domestic boiler) and

iv. Great fuel flexibility.

In fact, with this regard, it can be observed that the heat source of a TPV system can be provided by various fuel typologies such as fossil fuels (natural gas, oil, coke, etc.) municipal solid wastes, nuclear fuels, etc; concentrated solar radiation can also be used as a TPV heat source. A TPV system usually allows very low pollutant emissions (e.g. CO and NOx), since it is often coupled with combustion devices such as domestic boilers.

The main use of a TPV generator can be in the distributed combined heat and power generation, but its application in the automotive sector in case of hybrid vehicles, glass or other high temperatures industries has also been analyzed in literature. The TPV system has been proposed for portable generators, co-generation systems, combined cycle power plants, solar power plants, grid connected or independent equipment.

The electrical efficiency of the realized prototypes ranges from about 0.6 % to slightly less than 11.0 %. Moreover, electrical efficiencies close to 24 % are predicted in literature, making TPV system very attractive for cogeneration.

Electrical Performance of a TPV Generator

The power balance of a TPV generator is presented in below figure. The introduced power with fuel (P_{in}), unless the thermal losses ($P_{fuel,loss}$) of the combustion process, is converted by the emitter and by the optical filter into radiant power ($P'_{GAP} = P_{RAD} - P_{back}$) and thermal power discharged with the gases ($Q_{TH,gas}$ in figure and section F_K in figure). A fraction of the radiant power (P'_{GAP}), which is in the useful range of wavelengths for the photovoltaic conversion (due to the optical filter selection), can be lost due to the absorption of the optical filter (P_{abs}, even if this term can be usually neglected) and for the view factor between filter and PV cells (P_{loss} this term can be reduced achieving values very close to zero with a optimal design of the system geometry). The radiant power incident on the photovoltaic cells ($P_U = P_{GAP} - P_{loss} = P'_{GAP} - P_{loss} - P_{abs}$) is then converted into continuous current ($P_{el,dc}$) and thermal power ($Q_{th,pv}$); except for the losses ($P_{el,loss}$) due to the inverter (INV in figure) efficiency, the electrical power ($P_{el,ac}$) can be obtained from the system. On the other hand, the enthalpy content of the gases at the emitter exit ($_{TH\,gas}$ in figure and) can be partially recovered ($Q_{TH,cp}$) while the remaining part is discharged to the ambient ($Q_{th,d}$).

The electrical efficiency of a TPV generator can be written as:

$$\eta_{EL,TPV} = \eta_{CC} \cdot \eta_{RAD} \cdot \eta_{GAP} \cdot \eta_F \cdot \eta_{VF} \cdot \eta_{PV} \cdot \eta_{dc/ac},$$

Where: η_{CC}: combustion efficiency η_{RAD}: radiant efficiency; η_{GAP}: spectral efficiency; η_F: filter efficiency; η_{VF}: view factor efficiency; η_{PV}: cell efficiency $\eta_{dc/ac}$: inverter efficiency.

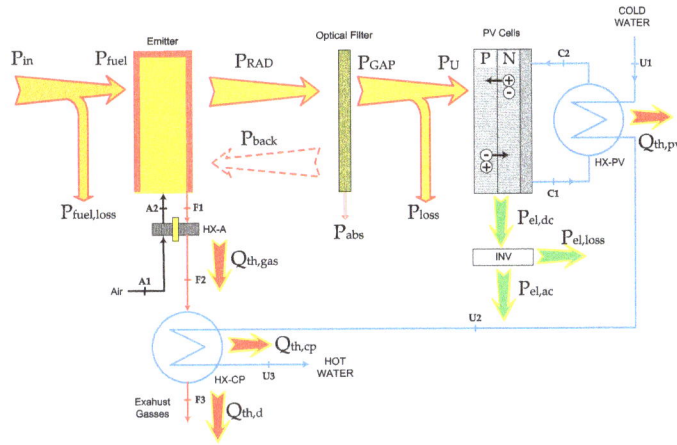

Figure: Schematics of a TPV generator

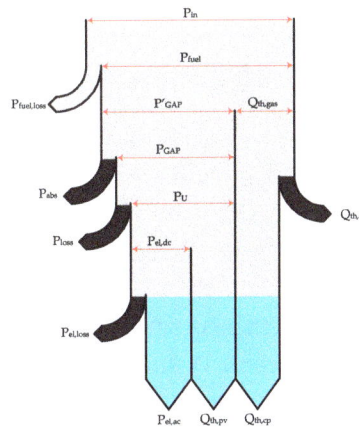

Figure: Power balance of a TPV generator

Combustion Efficiency

With reference to above in figure, the combustion efficiency can be expressed as the ratio between the useful introduced power (P_{fuel}) and the whole power introduced with fuel (P_{in}):

$$\eta_{CC} = \frac{P_{fuel}}{P_{in}} = \frac{P_{fuel}}{\dot{m}_{fuel}.LHV}$$

Being $\dot{m}_{fuel}$ and LHV respectively the fuel mass flow rate and its lower heating value (depending on the type of fuel).

It can be observed that the useful introduced power (P_{fuel}) can be only converted into radiant power ($P'_{GAP} = P_{RAD} - P_{back}$) or discharged with the combustion products ($Q_{TH,gas}$). It results:

$$P_{fuel} = P'_{GAP} - Q_{TH,gas}$$

This last equation represents the power balance of combustor, emitter and air pre-heater exchanger and allows to introduce the important role recovered by the air pre-heater for the thermophotovoltaic conversion. In facts, the reduction of combustion products temperature and then the increase of the air temperature upstream of the combustor allows to enhance the emitted radiant power; this evidence is also confirmed in. In particular, the air pre-heater was obtained by adopting a rotary heat exchanger with a ceramic material which is lighter than metal and with a greater heat capacity that is relevant to store an high value of energy; with this device an heat exchanger efficiency greater than 75% is achieved.

Radiant Efficiency

The radiant efficiency can be expressed, with reference to figure, as the ratio between the radiant power from the emitter (P_{RAD}) and the introduced power (P_{fuel}) into the system. It follows:

$$\eta_{RAD} = \frac{\eta_{RAD}}{P_{fuel}} = \frac{p_{RAD} \cdot S_{em}}{\eta_{CC} \cdot \dot{m}_{fuel} \cdot LHV}$$

The radiant power is a function of the radiation power density p_{RAD} and of the emitter surface S_{em}. The radiation efficiency is strictly influenced by several factors: the type of emitter, its dimension and thickness, the combustion mode, the firing rate and, the pre-heating of the air.

It can be observed that the radiant energy density can be up to 500 kW/m² (by integrating the spectrum of a black body at 1600°C). This last value is very high considering that the radiant energy density of the sun is equal to 1 kW/m² at AM1.5 condition. The achievement of high temperature is a very important aspect since, according to Planck's law, radiation power density scales with temperature to the fourth power. Therefore, most heat sources used in TPV systems are based on combustion systems; various types of premixed and non-premixed combustors or radiant tube burner have been developed in the last years. It should be considered that the aim toward high temperatures burner is limited by the NOx production. Anyway, recent studies about the emitter surface have shown that there exist an optimal value of emitter surface which allows the maximization of radiant efficiency and of emitter temperature been equal the boundary conditions (such as introduced power with fuel, emitter material, etc.) and the adopted equipment of the thermophotovoltaic generator.

The radiant energy from the emitter has to be characterized by an emission spectrum suitable for the adopted photovoltaic cells; in fact only the photon energy in a narrow band above the band gap of the photovoltaic cells can be converted into electrical energy. It follows that selective emission is required; in order to achieve this goal a selective emitter or a broadband emitter with a filter can be used. In the first case the emitter is made with materials such as rare earth oxides which are characterized by an emission spectrum centered on specific wavelength; in the second case, many of the emitted photons, due to their lower energy on the respect of the band gap of photovoltaic cells result unusable. It is imperative to send these photons back to the radiator in order to conserve heat and to reduce the fuel consumption needed to achieve the required emitter temperatures.

It should be observed that the material used for the emitter needs to have specific characteristics such as (i) thermal stability, (ii) corrosion resistance, (iii) shock resistance, (iv) high thermal conductivity, etc. Obviously, the high temperatures which are required by the TPV system implies that the emitter's material melting point should be as high as possible. Further, the emitter needs to be thermally stable in the selected atmosphere (i.e. air and/or combustion products) and high corrosion resistant; as example graphite (C) has a high-thermal conductivity and a good thermal shock but in an oxidizing atmosphere cannot overcome 400°C; on the contrary in non-oxidizing atmosphere can operate up to 3000°C. The adoption of coatings can improve the corrosion resistance of some materials or a shield, usually made of quartz, can be adopted to protect the emitter from the environment. A high value of thermal conductivity is required in order to have a uniform temperature distribution of the emitter. Lower values of thermal conductivities causes a large temperature gradient inside the emitter which drastically decreases its efficiency. Anyway, in case of porous emitter this factor may not be important. Thermal shock resistance is also very important especially in TPV generators with frequent on-off cycles. The sudden change in emitter temperature can cause material failure.

High temperature broadband emitters can be divided into (i) oxide based or (ii) non-oxide based ceramics. Among oxide based ceramics, alumina (Al_2O_3) and zirconia (ZrO_2) show a good stability in oxidizing atmosphere and can be used respectively up to 1900°C of temperature or more considering that their fusion temperatures are respectively 2050°C and 2600°C. Others oxide based ceramics are magnesia (MgO), silica (SiO_2), beryllia (BeO), hafnia (HfO_2), thoria (ThO_2) and yttria (Y_2O_3). Anyway, often the major difficult related to the adoption of these materials is the low thermal shock resistance and/or the low emissivity. A widely used broadband emitter is silicon carbide (SiC) which can operate up to 1650°C. It has an emissivity close to 0.90 and very high melting point. Ceramic composites such as SiC/Si and SiC coated ceramic composites fit all the requirements for a TPV emitter.

Spectral Efficiency

The spectral efficiency is the ratio between the whole radiation from the emitter (P_{RAD}) and the portion which pass through the filter (P'_{GAP}):

$$\eta_{GAP} = \frac{P'_{GAP}}{P_{RAD}}$$

The spectral efficiency depends on the adopted filter, used to match the emitter spectral emission to the PV cell; this means that the filter should be ideally able to block all the photons with energy lower than the PV cell band gap and pass the photons with higher energy. With a simple approach, the P'_{GAP} can be estimated by integrating the radiant intensity $I(\lambda;T_{em})$ in the range of wavelengths (from 0 to λ_{GAP}) which passes thought the filter and then can be converted by the photovoltaic cells:

$$P'_{GAP} = \varepsilon.\, S_{em} \int_0^{\lambda_{GAP}} I(\lambda;T_{em}).\tau(\lambda)d\lambda = \varepsilon.S_{em} \int_0^{\lambda_{GAP}} \frac{2\pi c^2}{\lambda^5}\left[\exp\left(\frac{hc}{\lambda k_B T_{em}}\right)-1\right]^{-1}.\tau(\lambda)d\lambda$$

Many types of filters have been developed such as plasma filters, 1-D photonic band gap filters, 2-D photonic band gap filters, 3-D photonic band gap filters, combination of plasma filter and 1-D photonic band gap filter, dielectric stacks or back-surface reflectors.

3-D photonic band gap filters are characterized by an omnidirectional photonic band gaps which means that the propagation of photons is prohibited for arbitrary polarization in any direction; obviously, this characteristic is highly appreciated for TPV generation. Anyway, it should be observed that a well designed 1-D photonic band gap filter can completely reflect polarized photons at all incident angles showing omnidirectional photonic band gaps. On this regards, filters based on multiple layer of SiO_2 have shown promising results for TPV applications.

Filter Efficiency

The filter efficiency takes into account the fraction of radiant power which is absorbed by the filter (P_{abs}) and which is lost. The filter efficiency can be written as:

$$\eta_F = \frac{P_{GAP}}{P'_{GAP}}$$

being the balance of the filter $\left(P_{RAD} - P_{back} = P'_{GAP} = P_{GAP} + P_{abs}\right)$. Usually the term P_{abs} can be neglected with a properly design of the filter and then it is possible to assume $\eta_F = 1$.

View Factor Efficiency

The view factor efficiency is related to the ratio between the radiation (P_U) which is incident on the photovoltaic cells and the value (P_{GAP}).

$$\eta_{VF} = \frac{P_U}{P_{GAP}}$$

The value of view factor can be calculated according to the geometry and to the distance among the surfaces which are involved in the irradiation phenomenon. Many formulations of radiation view factors can be found in literature on the basis of the TPV geometry.

PV Cells Efficiency

The cells efficiency represents the ratio between the electrical power output ($P_{el,dc}$) and the incident power on the cell (P_U); the maximum electrical power produced by a photovoltaic cell can be expressed as function of short circuit current (J_{SC}) open-circuit voltage (V_{OC}) and Fill Factor (FF). The radiation efficiency is influenced by many factors such as the cell material, the emitter temperature and the radiation intensity. It can be expressed as follows:

$$\eta_{PV} = \frac{P_{el,dc}}{P_U} = \frac{V_{OC} \cdot J_{SC} \cdot FF}{P_U}$$

Converters for TPV are very similar to standard solar cells such as *Si* and high efficiency *GaAs* but made of semiconductor materials with lower band gap, to get a better spectral matching with the emitter radiation.

A fundamental parameter in order to estimate the conversion efficiency of a PV cell is the external quantum efficiency $EQE(\lambda)$ which can be defined as the probability that a photon of wavelength

λ will be absorbed by the cell, generating an electron that will be collected at the terminals: it considers the reflection and absorption of incident photons and the generation/collection of minority carriers, so it describes the behavior of the *p-n* junction in great detail.

The actual value of J_{SC} produced by the cell can be calculated from EQE(λ) of the PV cell and the incident photon flow $\Phi(\lambda)$:

$$J_{SC} = e \int_{0}^{\lambda_{GAP}} \Phi(\lambda).EQE(\lambda).d(\lambda)$$

EQE(λ) were measured for different semiconductors of choice for TPV and typical behaviors are reported in figure. It could be noted that most of the materials used for the TPV cells have high *EQE* in a large region from near the band gap to lower wavelength. The *EQE* drops to very low value for photon wavelength of about 1000 nm, but it should be considered that in this region a standard TPV emitter at 1200-1800°C has a very low photon emission. For this reason, the TPV cells are usually able to convert with a very high efficiency the part of the black body radiation that arrives at their surface, while the photons with energy lower than the band gap, not being absorbed, can be effectively redirected towards the emitter with the use of appropriate selective filters. This particular characteristic, not possible for solar PV, permits TPV cells to potentially reach very high conversion efficiencies, because the incident radiation could be efficiently coupled to the region where the cell *EQE* is maximum.

Inverter Efficiency

Finally, the inverter efficiency allows the calculation of the final electrical output of the system. It results:

$$\eta_{dc/ac} = \frac{P_{el,ac}}{P_{el,dc}}$$

In particular, it can be observed that the use of transformer usually reduces the conversion efficiency from direct current to alternate current.

References

- Ignacio Luque-Heredia et al., "The Sun Tracker in Concentrator Photovoltaics" in Cristobal, A.B.,Martí, A.,and Luque, A. Next Generation Photovoltaics, Springer Verlag, 2012 ISBN 978-3642233692

- 10-disadvantages-and-advantages-of-solar-panels: connectusfund.org, Retrieved 25 June 2018

- David Lubitz William (2011). "Effect of Manual Tilt Adjustments on Incident Irradiance on Fixed and Tracking Solar Panels". Applied Energy. 88: 1710–1719. doi:10.1016/j.apenergy.2010.11.008.

- Different-types-solar-inverters: solarpowerworldonline.com, Retrieved 05 June 2018

- "Advanced Algorithm for MPPT Control of Photovoltaic System" (PDF). solarbuildings.ca. Archived from the original (PDF) on 2013-12-19. Retrieved 2013-12-19.

- Pros-and-cons-of-string-inverter-vs-microinverter: solarreviews.com, Retrieved 11 July 2018

- Surawdhaniwar, Sonali; Diwan, Ritesh (July 2012). "Study of Maximum Power Point Tracking Using Perturb and Observe Method". International Journal of Advanced Research in Computer Engineering & Technology. 1 (5): 106–110.

- String-inverters-advantages-disadvantages, learning-center: sinovoltaics.com, Retrieved 15 July 2018

- Antonio L. Luque; Viacheslav M. Andreev (2007). Concentrator Photovoltaics. Springer Verlag. ISBN 978-3-540-68796-2.

- Solar-wire-types-solar-pv-installations, support, installer: civicsolar.com, Retrieved 20 May 2018

- Seyedmahmoudian, Mehdi; Horan, Ben; Rahmani, Rasoul; Maung Than Oo, Aman; Stojcevski, Alex (2016-03-02). "Efficient Photovoltaic System Maximum Power Point Tracking Using a New Technique". Energies. 9 (3): 147. doi:10.3390/en9030147.

- Working-principle-of-a-solar-tracker: solartracker.guide, Retrieved 28 June 2018

- "Performances Improvement of Maximum Power Point Tracking Perturb and Observe Method". actapress.com. Retrieved 2011-06-18.

4

Applications of Photovoltaics

Photovoltaics is useful for a variety of applications like powering orbiting satellites and spacecraft, grid connected power generation, etc. The diverse applications of photovoltaics in the current scenario have been thoroughly discussed in this chapter, including the applications of photovoltaic appliances and transport applications.

Photovoltaic technology—or the use of solar energy to generate electricity—is one of the best break-throughs in the field of power generation.

There are a good number of applications for photovoltaic technology, including the following:

Photovoltaics for the Home

One of the places that will greatly benefit from photovoltaic systems is the home. Installing a solar power system in your home to augment your electrical supply will not be only cost-efficient, but will be a great step toward minimizing your carbon footprint and dependence on pollution-emitting technologies. Finding a small and affordable solar power system for your home is very easy, and they make for a quick and easy installation. Even the White House has installed its very own photovoltaic system to show Americans how cost-efficient and environment-friendly a solar-powered house is.

Photovoltaics for Business

Cutting overhead costs in order to maximize profits is a major goal for many companies. One of the biggest cost chunks in the overhead is electricity and power, which could be dramatically re-duced by installing photovoltaic solar panels. Not only will your profits be maximized, but you can also reduce your business' carbon footprint, enjoy energy independence, and save more money on equipment maintenance. Plus, taking steps to show that you care for the environment make for a good corporate social responsibility campaign.

Photovoltaics for the City

As of 2010, there are nearly 30 "megacities" in the world—these cities are not exactly the richest or the most advanced in the world, but they boast of over 10 million residents. The amount of homes and businesses in such cities is overwhelming—as are the electricity demands of the residents and workers in the city. There are many benefits of using solar power-generated electricity in the city: environmental, social, and economic.

Solar power does not need fuel in order to generate electricity, which means that there is no trace of waste or pollution emitted during the generation process. This also means that the sustainability of the city can be significantly improved by using photovoltaic technology. Solar panels can also be mounted on roofs, the sides of buildings, or basically any place where the panels can collect sun-light, without impact on the use of land or the ecology of certain city areas.

Photovoltaics can also increase employment in the city, and add to the business opportunities in the area. Investing in photovoltaic systems might feel like it is costly in the beginning, but would be a worthwhile investment for both individuals and businesses in the long run.

Photovoltaics in Sport

Applying the technology of photovoltaics in the world of sport is not a quick leap, but rest assured that clean and sustainable energy is being embraced by this field. In recent times we've seen the emergence of solar power chargers for uses in sport and everyday life. These items are a clever way to extend the life of popular items such as smartphones, GPS equipment and any battery operated chargeable item. By simply using the sun you can effectively charge any of these devices. These items could prove to be essential emergency backups especially when hiking.

Many new stadiums—especially those ones built for much awaited sporting events such as the Summer Olympics and the FIFA World Cup—are using photovoltaic technology to power the whole structure. Aside from dramatically reducing the operating costs of the stadiums, the move toward more sustainable and renewable energies in the world of sports is adding another integral value to its credo: the conscious use of clean energy for a better future.

Appliances Applications

Solar-powered Refrigerator

Solar-powered refrigerator is a cooling appliance that works completely with energy provided by the sun. They are mostly used in hot climates to keep food so it would not perish and to keep vaccines at their appropriate temperature. Some solar-powered refrigerators use combination of solar panels and lead batteries to store energy when there is no sun (on the cloudy days and at night), while other use heat of the sun to make refrigerator work.

The first to have an idea for a solar-powered refrigerator was an engineer Otto Mohr in 1935. He designed all of the correct components needed to build the solar refrigerator and with them it would work. Since then many engineers, scientists, and researchers improved his ideas and built their working variants of a solar-powered refrigerator. In 1950's USA installed a flat plate collector system. At the same time USSR made a parabolic mirror which could produce 250kg of ice per day. An absorption machine with a cylinder-parabolic mirror was made in France and it could made 100kg of ice per day. There were also built systems that charged during the day and cooled during the night. Today, solar refrigerators are mostly used for camping and in developing countries where electricity is scarce.

Refrigerators generally keep food cool through the process of evaporation. Compressor, which is a part of a refrigerator, turns the refrigerant gas into a liquid. Pressure is then removed; the refrigerant turns back into a gas and absorbs heat because it has to take energy from somewhere. That cools its surroundings. Solar-powered refrigerator works similarly but the refrigerant is replaced with ammonia or lithium bromide mixed with water. Instead of compressor, the heat of the sun is used to increase the pressure of the gas. From here solar refrigerator works as a common refrigerator: gas again turns into a liquid, heat is removed, the liquid evaporates and lowers the temperature.

Solar-powered refrigerator is useful for many reasons. First it is used in regions that have no electricity but have greater number of sunny days a year. Secondly, such regions often use kerosene or gas-powered absorption refrigerated coolers which require constant supply of fuel and generate large amounts of carbon dioxide. They are also very difficult to adjust which can result in the freezing of medicine which is kept in those coolers. Fuel is also costly and there is risk of causing fires. Solar-powered refrigerator can replace these kinds of coolers and make cooling safer.

Solar-powered refrigerator that have batteries work on electricity but it is supplied from solar panels. These refrigerators are expensive and have heavy lead-acid batteries which deteriorate quickly in hot weather. Batteries also must be replaced approximately every three years which makes them more costly and there is also disposing of these batteries because, if not disposed properly, they can result in lead pollution. Solar-powered refrigerators don't have these problems but they are very dependent on the sun and on the cloudy days they don't work and have to have reserve systems that will maintain low temperatures on such occasions.

Types of Solar Refrigeration System

Solar refrigeration systems can be classified in three different categories. They are:

    i. Photovoltaic operated refrigeration system

    ii. Solar mechanical refrigeration

    iii. Absorption refrigeration

Photovoltaic Operated Refrigeration System

In this system, solar radiation is directly converted to direct current electricity using semiconducting materials. The operation of a PV-powered solar refrigeration cycle is simple. Solar photovoltaic panels produce DC electrical power that can be used to operate a DC motor, which is coupled to the compressor of a vapour compression refrigeration system. The process that makes the refrigeration possible is the conversion of sunlight into DC electrical power, achieved by the PV panel. The DC electrical power drives the compressor to circulate refrigerant through a vapour compression refrigeration loop that extracts heat from an insulated enclosure. This enclosure includes the thermal reservoir and a phase change material. Figure shows the schematic diagram of a photovoltaic operated refrigeration system.

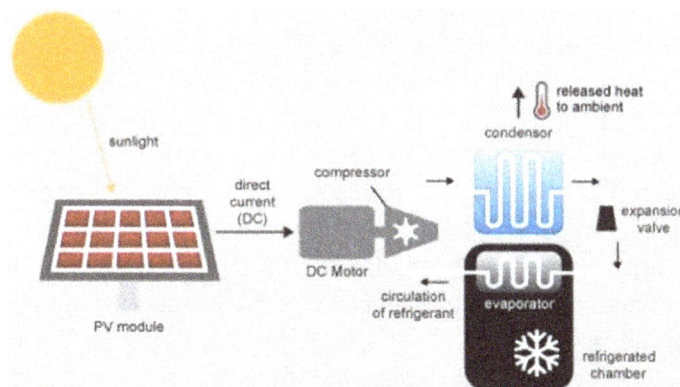

Figure: Schematic diagram of a photovoltaic operated refrigeration system

Solar Mechanical Refrigeration

In this type of refrigeration system, required compressor power to drive the compressor in refrigeration cycle is provided by a solar Rankine cycle. Sunlight strikes the solar panel which drives a Rankine cycle and produces work in the turbine. This work is then utilized to run the compressor of the vapour compression refrigeration system. The schematic diagram of a solar mechanical refrigeration system has been shown in figure.

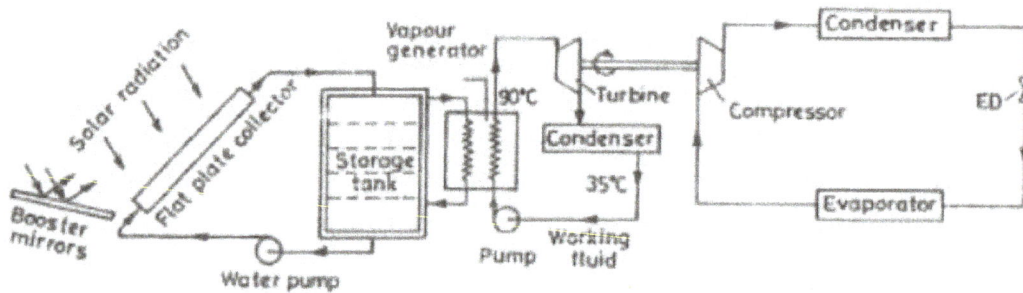

Figure: Schematic diagram of a solar mechanical vapour compression refrigeration system

Solar Absorption Refrigeration System

In this system, low grade energy as heat from solar panel is used as input for chilling purpose.

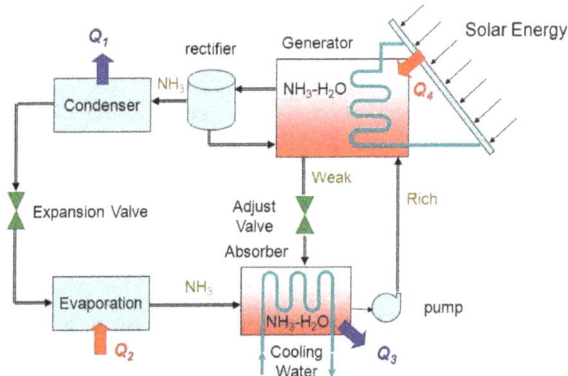

Figure: Schematic diagram of a solar absorption refrigeration system

Figure shows the schematic diagram of a solar absorption refrigeration system. This system is different from a conventional vapour compression refrigeration system. Basic components of such refrigeration system are absorber, generator, solar panel, condenser, expansion valve, evaporator, DC battery and fan. The compressor in the vapour compression system is replaced by a generator, absorber and pump. Refrigerant (NH_3) in the evaporator absorbs the heat from the refrigerated space and gets evaporated. It is then passed to absorber where it is dissolved with absorbent (H_2O) and pumped to generator. Electrical energy from solar panel is utilized for heating in the generator and the refrigerant enters into condenser. The refrigerant is converted to liquid in the condenser and the pressure of the liquid refrigerant is dropped to the evaporator pressure with the help of an expansion device (ED). The main advantage of absorption system is compression of liquid instead of vapour which results in less mechanical work requirement as input. But the system is much expansive compared to compression refrigeration system. Other than ammonia-water combination, few more refrigerant - absorber pairs have been tried which are listed in table.

Table: Different refrigerant – absorber pairs

Refrigerant	Absorber	State of Absorber
Ammonia	Sodium thiocynate	Solid
Ammonia	Lithium Nitrate	Solid
Ammonia	Calcium Chloride	Solid
Ammonia	Isobutene	Solid
Water	Lithium Bromide	Solid
Water	Lithium Chloride	Solid
Methyl Chloride	Dimethyl Ether or Tetra Ethylene Glycol	Liquid

The performance of a refrigeration system is judged by a parameter called coefficient of performance (COP). The COP of a solar absorption refrigeration system can be expressed as

$$COP = \frac{\text{Heat absorbed in the evaporator}}{\text{Generator heat supply} + \text{Pump work input}}$$

As pump work is very small and the above expression can be approximated without much error as

$$COP = \frac{\text{Heat absorbed in the evaporator}}{\text{Generator heat supply}} = \frac{Q_2}{Q_4}$$

General energy equation for flat plate collector is given by,

$$Q = I \times A_C$$

where,

Q = Amount of solar radiation received by collector =

$$\dot{m} \times c_p \times \left(T_p - T_a\right)$$

I = Intensity of solar radiation

and AC = Collector surface area

Useful heat gain by the collector can be written as

$$Q_u = F_R \times A_C \times \left(S - U_1 \times \left(T_p - T_a\right)\right)$$

where, S = Incident solar flux absorbed by the absorber plate

F_R = Collector heat removal factor

U_1 = Overall heat loss co-efficient

T_p = Fluid temperature at inlet of collector

Solar Absorption Refrigeration System Performance

Different researchers investigated on absorption refrigeration system varying different parameters and recorded the performance of the system under such varying conditions. Karno and Ajib simulated a vapour absorption refrigeration system using acetone-zinc bromide solutions and reported that initially the COP of the system increased rapidly then the increment was found to be slightly flatter in nature with the increase in generator temperature for fixed evaporator temperature. The results are shown for condenser and absorber temperature both fixed at 28°C and refrigerant heat exchanger effectiveness at 75%. However, COP of the system increases with the increase in evaporator temperature when other parameters are kept constant. An approximate correlation for COP of the system in terms of generator temperature (TG) and evaporator temperature (TE) has also been developed which is as follows:

$$\text{COP} = \frac{0.78 - 0.0134 \times T_G - 0.015 \times T_E}{1 - 0.018 \times T_G - 0.0177 \times T_E}$$

The variations of COP with generator temperature and evaporator temperature have been shown in below figure. Calculated values from the simulation program have been plotted together with the values calculated from the above said correlation.

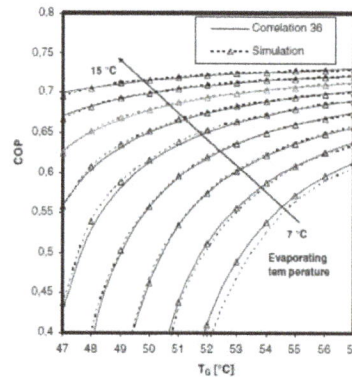

Figure: Variation of COP with generator temperature and evaporator temperature

Effects of evaporator temperature at different condenser or absorber temperature on COP have also been shown in below figure. The generator temperature is kept constant at 57°C. It shows that the COP of the system decreases with the increase in condenser or absorber temperature for any evaporator temperature.

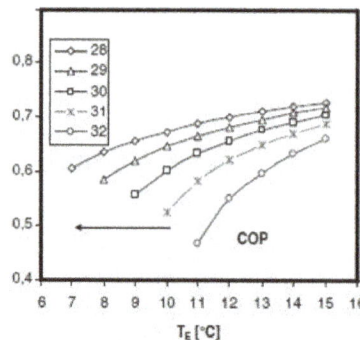

Figure: Effect of evaporator temperature on COP for different condenser/absorber temperature (18-32°C)

Advantages of Solar Refrigeration

Solar energy is the main source of energy that is utilized to run solar refrigerator. So, significant amount of electrical power is saved and it also causes less pollution that would have been added due to the use of power produced by the conventional power plants. The solar energy is available in every part of the world and unlike fossil fuels and nuclear power; it is a clean source of energy. Additional power from the solar collector can also be used for the other domestic purposes. The solar refrigerators can be very useful where there is no continuous supply of electricity or difficult to get conventional fuel. More importantly it is renewable in nature. Conventional refrigeration systems emit significant amount of gas which pollute the environment. This solar refrigeration system is also needed to lower the environmental impact caused due to conventional refrigeration systems. The maintenance cost of such system is considerably low compared to that of the conventional system. Those facts encourage to use solar refrigeration system whenever possible.

Disadvantages of Solar Refrigeration

Solar refrigeration systems also have some disadvantages. As solar radiation is not available throughout the day, power production is not uniform. Again it depends on the intensity of the beam radiation. Even in the hottest regions on earth, the average solar radiation flux rarely exceeds 1 kWh/m² and the maximum radiation flux over a day is about 6 kWh/m². These are low values from the point of view of technological utilization. So, those refrigeration systems can be used in those places where those problems are not present. To produce sufficient energy from solar system, it needs bigger collector. So, there is a need of bigger space for the collector which is another major problem for using solar refrigeration system. Initial investment to develop such set up is also large.

Challenges

The variation in availability of solar radiation occurs daily because of the day-night cycle and also seasonally because of the earth's orbital motion around the sun. In addition, variation occurs at a specific location because of local weather conditions. Consequently, the energy collected when the sun is shining must be stored for use during periods when it is not available. The need for storage significantly adds to the cost of the system. Thus, the real challenge in utilizing solar energy as an energy alternative is to address these challenges. One has to strive for the development of cheaper methods of collection and storage so that the large initial investments required at present in most applications are reduced.

Solar Air Conditioning

A solar air conditioning system is simply a system of cooling and heating that utilises solar power, rather than electricity from the mains. While they can have a significant upfront installation cost, solar air conditioners are much cheaper to own and operate in the long term, as they utilise virtually unlimited energy provided by our sun.

Solar energy is a great way to decrease both your electricity bill and your carbon footprint, so what are some different ways to incorporate it into your home?

- Conventional solar powered cooling systems, which comprise the majority of installed systems.

- 'Open loop' cooling systems (known as desiccant cooling). This type of cooling is newer and less common, but can be more cost and energy efficient.

Conventional solar air conditioners utilize a standard vapor compression refrigeration cycle to cool or heat the building in question. The essential principle is that heat in the building is transferred to a refrigerant gas, which is then circulated out to a condenser where the excess heat is dumped. The now cool refrigerant then circulates back into the building to repeat the cycle. A conventional solar AC system draws operational power from a solar array rather than the electricity grid, in an effort to increase efficiency and lower costs.

In this instance, the solar energy heats a hot water tank, and this water is then used in two ways as part of the air conditioning system:

- When cooling is required, the water is used to heat up air, which is in turn used to dry out a cylinder that extracts the excess moisture from air flowing into a parallel chamber. This dried air is then pumped through a cooler and is then circulated around the house. The hot air used for drying is then exhausted outside the building in this cooling process.

- When heating is required, the hot air is instead used to directly warm the house.

Photovoltaic (PV) Solar Cooling

Photovoltaics can provide the power for any type of electrically powered cooling be it conventional compressor-based or adsorption/absorption-based, though the most common implementation is with compressors. For small residential and small commercial cooling (less than 5 MWh/a) PV-powered cooling has been the most frequently implemented solar cooling technology. The reason for this is debated, but commonly suggested reasons include incentive structuring, lack of residential-sized equipment for other solar-cooling technologies, the advent of more efficient electrical coolers, or ease of installation compared to other solar-cooling technologies (like radiant cooling).

Since PV cooling's cost effectiveness depends largely on the cooling equipment and given the poor efficiencies in electrical cooling methods until recently it has not been cost effective without subsidies. Using more efficient electrical cooling methods and allowing longer payback schedules is changing that scenario.

For example, a 100,000 BTU U.S. Energy Star rated air conditioner with a high seasonal energy efficiency ratio (SEER) of 14 requires around 7 kW of electric power for full cooling output on a hot day. This would require over a 20 kW solar photovoltaic electricity generation system with storage.

A solar-tracking 7 kW photovoltaic system would probably have an installed price well over $20,000 USD (with PV equipment prices currently falling at roughly 17% per year). Infrastructure, wiring, mounting, and NEC code costs may add up to an additional cost; for instance a 3120 watt solar panel grid tie system has a panel cost of $0.99/watt peak, but still costs ~$2.2/watt hour peak. Other systems of different capacity cost even more, let alone battery backup systems, which cost even more.

A more efficient air conditioning system would require a smaller, less-expensive photovoltaic system. A high-quality geothermal heat pump installation can have a SEER in the range of 20 (±). A 100,000 BTU SEER 20 air conditioner would require less than 5 kW while operating.

Newer and lower power technology including reverse inverter DC heat pumps can achieve SEER ratings up to 26.

There are new non-compressor-based electrical air conditioning systems with a SEER above 20 coming on the market. New versions of phase-change indirect evaporative coolers use nothing but a fan and a supply of water to cool buildings without adding extra interior humidity (such as at McCarran Airport Las Vegas Nevada). In dry arid climates with relative humidity below 45% (about 40% of the continental U.S.) indirect evaporative coolers can achieve a SEER above 20, and up to SEER 40. A 100,000 BTU indirect evaporative cooler would only need enough photovoltaic power for the circulation fan (plus a water supply).

A less-expensive partial-power photovoltaic system can reduce (but not eliminate) the monthly amount of electricity purchased from the power grid for air conditioning (and other uses). With American state government subsidies of $2.50 to $5.00 USD per photovoltaic watt, the amortized cost of PV-generated electricity can be below $0.15 per kWh. This is currently cost effective in some areas where power company electricity is now $0.15 or more. Excess PV power generated when air conditioning is not required can be sold to the power grid in many locations, which can reduce (or eliminate) annual net electricity purchase requirement.

Superior energy efficiency can be designed into new construction (or retrofitted to existing buildings). Since the U.S. Department of Energy was created in 1977, their Weatherization Assistance Program has reduced heating-and-cooling load on 5.5 million low-income affordable homes an average of 31%. A hundred million American buildings still need improved weatherization. Careless conventional construction practices are still producing inefficient new buildings that need weatherization when they are first occupied.

It is fairly simple to reduce the heating-and-cooling requirement for new construction by one half. This can often be done at no additional net cost, since there are cost savings for smaller air conditioning systems and other benefits.

Geothermal Cooling

Earth sheltering or Earth cooling tubes can take advantage of the ambient temperature of the Earth to reduce or eliminate conventional air conditioning requirements. In many climates where the majority of humans live, they can greatly reduce the buildup of undesirable summer heat, and also help remove heat from the interior of the building. They increase construction cost, but reduce or eliminate the cost of conventional air conditioning equipment.

Earth cooling tubes are not cost effective in hot humid tropical environments where the ambient Earth temperature approaches human temperature comfort zone. A solar chimney or photovoltaic-powered fan can be used to exhaust undesired heat and draw in cooler, dehumidified air that has passed by ambient Earth temperature surfaces. Control of humidity and condensation are important design issues.

A geothermal heat pump uses ambient Earth temperature to improve SEER for heat and cooling. A deep well recirculates water to extract ambient Earth temperature (typically at 2 gallons of water per ton per minute). These "open loop" systems were the most common in early systems, however water quality could cause damage to the coils in the heat pump and shorten the life of the equipment. Another method is a closed loop system, in which a loop of tubing is run down a well or wells, or in trenches in the lawn, to cool an intermediate fluid. When wells are used, they are backfilled with Bentonite or another grout material to ensure good thermal conductivity to the earth.

In the past, the fluid of choice was a 50/50 mixture of propylene glycol because it is non-toxic unlike ethylene glycol (which is used in car radiators). Propylene glycol is viscous, and would eventually gum up some parts in the loop(s), so it has fallen out of favor. Today, the most common transfer agent is a mixture of water and ethyl alcohol (ethanol).

Ambient earth temperature is much lower than peak summer air temperature, and much higher than the lowest extreme winter air temperature. Water is 25 times more thermally conductive than air, so it is much more efficient than an outside air heat pump, (which becomes less effective when the outside temperature drops in winter).

The same type of geothermal well can be used without a heat pump but with greatly diminished results. Ambient Earth temperature water is pumped through a shrouded radiator (like an automobile radiator). Air is blown across the radiator, which cools without a compressor-based air conditioner. Photovoltaic solar electric panels produce electricity for the water pump and fan, eliminating conventional air-conditioning utility bills. This concept is cost-effective, as long as the location has ambient Earth temperature below the human thermal comfort zone (not the tropics).

Solar Open-loop Air Conditioning using Desiccants

Air can be passed over common, solid desiccants (like silica gel or zeolite) or liquid desiccants (like lithium bromide/chloride) to draw moisture from the air to allow an efficient mechanical or evaporative cooling cycle. The desiccant is then regenerated by using solar thermal energy to dehumidfy, in a cost-effective, low-energy-consumption, continuously repeating cycle. A photovoltaic system can power a low-energy air circulation fan, and a motor to slowly rotate a large disk filled with desiccant.

Energy recovery ventilation systems provide a controlled way of ventilating a home while minimizing energy loss. Air is passed through an "enthalpy wheel" (often using silica gel) to reduce the cost of heating ventilated air in the winter by transferring heat from the warm inside air being exhausted to the fresh (but cold) supply air. In the summer, the inside air cools the warmer incoming supply air to reduce ventilation cooling costs. This low-energy fan-and-motor ventilation system can be cost-effectively powered by photovoltaics, with enhanced natural convection exhaust up a solar chimney - the downward incoming air flow would be forced convection (advection).

A desiccant like calcium chloride can be mixed with water to create an attractive recirculating waterfall, that dehumidifies a room using solar thermal energy to regenerate the liquid, and a PV-powered low-rate water pump.

Active solar cooling wherein solar thermal collectors provide input energy for a desiccant cooling system. There are several commercially available systems that blow air through a desiccant impregnated medium for both the dehumidification and the regeneration cycle. The solar heat is one way that the regeneration cycle is powered. In theory, packed towers can be used to form a counter-current flow of the air and the liquid desiccant but are not normally employed in commercially available machines. Preheating of the air is shown to greatly enhance desiccant regeneration. The packed column yields good results as a dehumidifier/regenerator, provided pressure drop can be reduced with the use of suitable packing.

Passive Solar Cooling

In this type of cooling solar thermal energy is not used directly to create a cold environment or drive any direct cooling processes. Instead, solar building design aims at slowing the rate of heat transfer into a building in the summer, and improving the removal of unwanted heat. It involves a good understanding of the mechanisms of heat transfer: heat conduction, convective heat transfer, and thermal radiation, the latter primarily from the sun.

For example, a sign of poor thermal design is an attic that gets hotter in summer than the peak outside air temperature. This can be significantly reduced or eliminated with a cool roof or a green roof, which can reduce the roof surface temperature by 70°F (40°C) in summer. A radiant barrier and an air gap below the roof will block about 97% of downward radiation from roof cladding heated by the sun.

Passive solar cooling is much easier to achieve in new construction than by adapting existing buildings. There are many design specifics involved in passive solar cooling. It is a primary element of designing a zero energy building in a hot climate.

Solar Closed-loop Absorption Cooling

The following are common technologies in use for solar thermal closed-loop air conditioning:

- Absorption: NH_3/H_2O or Ammonia/Water,
- Absorption: Water/Lithium Bromide,
- Absorption: Water/Lithium Chloride,
- Adsorption: Water/Silica Gel or Water/Zeolite,
- Adsorption: Methanol/Activated Carbon.

Active solar cooling uses solar thermal collectors to provide solar energy to thermally driven chillers (usually adsorption or absorption chillers). Solar energy heats a fluid that provides heat to the generator of an absorption chiller and is recirculated back to the collectors. The heat provided to the generator drives a cooling cycle that produces chilled water. The chilled water produced is used for large commercial and industrial cooling.

Solar thermal energy can be used to efficiently cool in the summer, and also heat domestic hot water and buildings in the winter. Single, double or triple iterative absorption cooling cycles are used in different solar-thermal-cooling system designs. The more cycles, the more efficient they are. Absorption chillers operate with less noise and vibration than compressor-based chillers, but their capital costs are relatively high.

Efficient absorption chillers nominally require water of at least 190°F (88°C). Common, inexpensive flat-plate solar thermal collectors only produce about 160°F (71°C) water. High temperature flat plate, concentrating (CSP) or evacuated tube collectors are needed to produce the higher temperature transfer fluids required. In large scale installations there are several projects successful both technical and economical in operation worldwide including, for example, at the headquarters of Caixa Geral de Depósitos in Lisbon with 1,579 square metres (17,000 sq ft) solar collectors and 545 kW cooling power or on the Olympic Sailing Village in Qingdao/China. In 2011 the most powerful plant at Singapore's new constructed United World College will be commissioned (1500 kW).

These projects have shown that flat plate solar collectors specially developed for temperatures over 200°F (93°C) (featuring double glazing, increased backside insulation, etc.) can be effective and cost efficient. Where water can be heated well above 190°F (88°C), it can be stored and used when the sun is not shining.

The Audubon Environmental Center at the Ernest E. Debs Regional Park in Los Angeles has an example solar air conditioning installation, which failed fairly soon after commissioning and is no longer being maintained. The Southern California Gas Co. (The Gas Company) is also testing the practicality of solar thermal cooling systems at their Energy Resource Center (ERC) in Downey, California. Solar Collectors from Sopogy and Cogenra were installed on the rooftop at the ERC and are producing cooling for the building's air conditioning system. Masdar City in the United Arab Emirates is also testing a double-effect absorption cooling plant using Sopogy parabolic trough collectors, Mirroxx Fresnel array and TVP Solar high-vacuum solar thermal panels.

For 150 years, absorption chillers have been used to make ice (before the electric light bulbs were invented). This ice can be stored and used as an "ice battery" for cooling when the sun is not shining, as it was in the 1995 Hotel New Otani Tokyo in Japan. Mathematical models are available in the public domain for ice-based thermal energy storage performance calculations.

The ISAAC Solar Icemaker is an intermittent solar ammonia-water absorption cycle. The ISAAC uses a parabolic trough solar collector and a compact and efficient design to produce ice with no fuel or electric input, and with no moving parts.

Providers of solar cooling systems include ChillSolar, SOLID, Sopogy, Cogenra, Mirroxx and TVP Solar for commercial installations and ClimateWell, Fagor-Rotartica, SorTech and Daikin mostly for residential systems. Cogenra uses solar co-generation to produce both thermal and electric energy that can be used for cooling.

Solar Cooling Systems utilizing Concentrating Collectors

The main reasons for employing concentrating collectors in solar cooling systems are: high efficient air-conditioning through coupling with double/triple effect chillers; and solar refrigeration serving industrial end-users, possibly in combination with process heat and steam.

Concerning industrial applications, several studies in the recent years highlighted that there is a high potential for refrigeration (temperatures below 0°C) in different areas of the globe (e.g., the Mediterranean, Central America). However, this can be achieved by ammonia/ water absorption chillers requiring high temperature heat input at the generator, in a range (120 ÷ 180°C) which can only be satisfied by concentrating solar collectors. Moreover, several industrial applications require both cooling and steam for processes, and concentrating solar collectors can be very advantageous in the sense that their use is maximized.

Zero-energy Buildings

Goals of zero-energy buildings include sustainable, green building technologies that can significantly reduce, or eliminate, net annual energy bills. The supreme achievement is the totally off-the-grid autonomous building that does not have to be connected to utility companies. In hot climates with significant degree days of cooling requirement, leading-edge solar air conditioning will be an increasingly important critical success factor.

Solar Lamp

A solar lamp in Rizal Park, Philippines   A garden solar lamp

A solar lamp also known as solar light or solar lantern, is a lighting system composed of an LED lamp, solar panels, battery, charge controller and there may also be an inverter. The lamp operates on electricity from batteries, charged through the use of solar photovoltaic panel.

Magnifying glass

Solar-powered household lighting can replace other light sources like candles or kerosene lamps. Solar lamps have a lower operating cost than kerosene lamps because renewable energy from the sun is free, unlike fuel. In addition, solar lamps produce no indoor air pollution unlike kerosene lamps. However, solar lamps generally have a higher initial cost, and are weather dependent.

Solar lamps for use in rural situations often have the capability of providing a supply of electricity for other devices, such as for charging cell phones. American investors have been working towards developing a $10 / unit solar lantern for replacement of kerosene lamps.

Components

The whole structure of solar lamp is shown in figure.

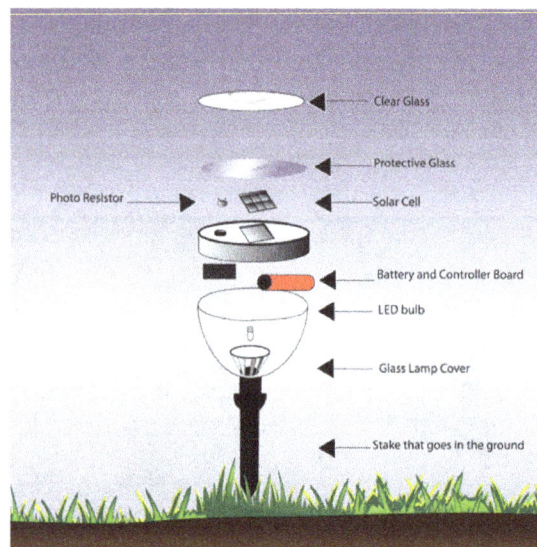

Figure: Structure of solar lamp

Solar Panels

Solar panels are made out of crystals that are made out of covalent bonds between electrons on the outer shell of silicon atoms. Silicon is a semiconductor which is neither metals that conducts electricity nor insulators that do not conduct electricity. Semiconductors normally do not conduct electricity but under certain circumstances they do in this example with exposition to light.

A solar cell has two different layers of silicon. The lower layer has less electrons and hence has a slight positive charge due to the negative charge nature of electrons. In addition, the upper layer has more electrons and has slightly negative charge.

A barrier is created between these two layers however when the stream of light particles called photons enter, they give up their energy to the atoms in the silicon. It promotes one electron from a covalent bond to a next energy level from upper layer to the lower layer. This promotion of an electron allows freer movement within the crystal which produces a current. More light shines through, more electrons move around hence more current flows between. This process is called photovoltaic and photoelectric effect. Photovoltaic systems literally means combination of light and voltage and they use photovoltaic cells to directly convert sunlight into electricity.

Solar panels are made out of layers of different materials, in figure in order of glass, encapsulate, crystalline cells, encapsulate, back sheet, junction box and lastly frame. The encapsulate keeps out moisture and contaminants which could cause problems.

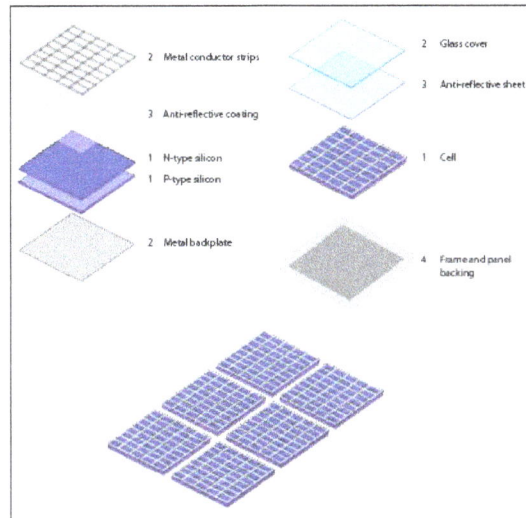

Figure

Battery

A battery is usually housed within a metal or plastic case. Inside the case are electrodes including cathodes and anodes where chemical reactions occur. A separator also exists between cathode and anode which stops the electrodes reacting together at the same time as allowing electrical charge to flow freely between the two. Lastly, the collector conducts a charge from the battery to outside.

Batteries inside solar lamps usually use gel electrolyte technology with high performance in deep discharging, in order to enable use in extreme ranges of temperature. It may also use lead-acid, nickel metal hydride, nickel cadmium, or lithium.

This part of the lamp saves up energy from the solar panel and provides power when needed at night when there is no light energy available.

In general, the efficiency of photovoltaic energy conversion is limited for physical reasons. Around 24% of solar radiation of a long wavelength is not absorbed. 33% is heat lost to surroundings, and further losses are of approximately 15-20%. Only 23% is absorbed which means a battery is a crucial part of solar lamp.

Charge Controller

This section controls the entire working systems to protect battery charge. It ensures, under any circumstances including extreme weather conditions with large temperature difference, the battery does not overcharge or over discharge and damage the battery even further.

This section also includes additional parts such as light controller, time controller, sound, temperature compensation, lighting protection, reverse polarity protection and AC transfer switches which ensure sensitive back-up loads work normally when outage occurs.

Working Principles

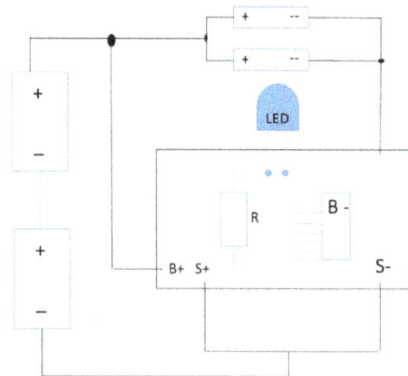

LED lights are used due to their high luminous efficiency and long life. Under the control of a DC charge controller, non-contact control automatically turns on the light at dark and switches off at daytime. It sometimes also combines with time controllers to set curtain time for it to automatically switch light on and off.

As shown in figure, the chip includes microchip(R), B-, B+, S- and S+. S+ and S- are both connected to solar panels with wire, one of which has plus charge and the other minus charge. B- and B+ are attached to two batteries in this case. The light will be shown through the LED light when all of these are connected.

Benefits

Solar lamps can be easier for customers to install and maintain as they do not require an electricity cable. Solar lamps can benefit owners with reduced maintenance cost and costs of electricity bills. Solar lamps can also be used in areas where there is no electrical grid or remote areas that lack a reliable electricity supply. There are many stories of people with lung disease, eye deterioration, burns and sometimes even death simply because they do not have a healthy alternative to light at night. Women have felt unsafe walking to the toilet outside after dark. Babies are being delivered by midwives using only a candle, and students cannot study when the sun goes down for lack of light leading to increased illiteracy and perpetual poverty. These are the realities for over 1 billion people around the globe. Lack of lighting equates to continued poverty felt around the world.

Solar energy output is limited by weather and can be less effective if it is cloudy, wet, or winter.

Households switching to solar lamps from kerosene lamps also gain from health risk associated with kerosene emissions. Kerosene often has negative impacts on human lungs.

The use of solar energy minimizes the creation pollution indoors, where kerosene have been linked to cases of health issues. However, photovoltaic panels are made out of silicon and other toxic metals including lead that can be difficult to dispose of.

The use of solar lights improves education for students who live in households without electricity. When the nonprofit Unite-To-Light donated solar-lamps to schools a remote region of Kwa Zulu Natal in South Africa test scores and pass rates improved by over 30%. The light gives students added time to study after dark.

A 2017 experimental study in unelectrified areas of northern Bangladesh found that the use of solar lanterns decreased total household expenditure, increased children's home-study hours and increased school attendance. It did not however improve the children's educational achievement to any large extent.

Use

Solar Street Light

These lights provide a convenient and cost-effective way to light streets at night without the need of AC electrical grids for pedestrians and drivers. They may have individual panels for each lamp of a system, or may have a large central solar panel and battery bank to power multiple lamps.

Solar LED lantern

Rural

In rural India, solar lamps, commonly called solar lanterns, using either LEDs or CFLs, are being used to replace kerosene lamps. Especially in areas where electricity is otherwise difficult to access, solar lamps are very useful and it will also improve life in rural areas.

Solar Charge Controller

A solar charge controller is fundamentally a voltage or current controller to charge the battery and keep electric cells from overcharging. It directs the voltage and current hailing from the solar panels setting off to the electric cell. Generally, 12V boards/panels put out in the ballpark of 16 to 20V, so if there is no regulation the electric cells will damaged from overcharging. Generally, electric storage devices require around 14 to 14.5V to get completely charged. The solar charge controllers are available in all features, costs and sizes. The range of charge controllers are from 4.5A and up to 60 to 80A.

Types of Solar Charger Controller

There are three different types of solar charge controllers, they are:

1. Simple 1 or 2 stage controls

2. PWM (pulse width modulated)

3. Maximum power point tracking (MPPT)

Simple 1 or 2 Controls: It has shunt transistors to control the voltage in one or two steps. This controller basically just shorts the solar panel when a certain voltage is arrived at. Their main genuine fuel for keeping such a notorious reputation is their unwavering quality – they have so not many segments, there is very little to break.

PWM (pulse width modulated): This is the traditional type charge controller, for instance anthrax, Blue Sky and so on. These are essentially the industry standard now.

Maximum power point tracking (MPPT): The MPPT solar charge controller is the sparkling star of today's solar systems. These controllers truly identify the best working voltage and amperage of the solar panel exhibit and match that with the electric cell bank. The outcome is extra 10-30% more power out of your sun oriented cluster versus a PWM controller. It is usually worth the speculation for any solar electric systems over 200 watts.

Features of Solar Charge Controller

- Protects the battery (12V) from over charging,

- Reduces system maintenance and increases battery lifetime,

- Auto charged indication,

- Reliability is high,

- 10amp to 40amp of charging current,

- Monitors the reverse current flow.

Function of Solar Charge Controller

The most essential charge controller basically controls the device voltage and opens the circuit, halting the charging, when the battery voltage ascents to a certain level. More charge controllers utilized a mechanical relay to open or shut the circuit, halting or beginning power heading off to the electric storage devices.

Generally solar power systems utilize 12V of batteries. Solar panels can convey much more voltage than is obliged to charge the battery. The charge voltage could be kept at a best level while the time needed to completely charge the electric storage devices is lessened. This permits the solar systems to work optimally constantly. By running higher voltage in the wires from the solar panels to the charge controller, power dissipation in the wires is diminished fundamentally.

The solar charge controllers can also control the reverse power flow. The charge controllers can distinguish when no power is originating from the solar panels and open the circuit separating the solar panels from the battery devices and halting the reverse current flow.

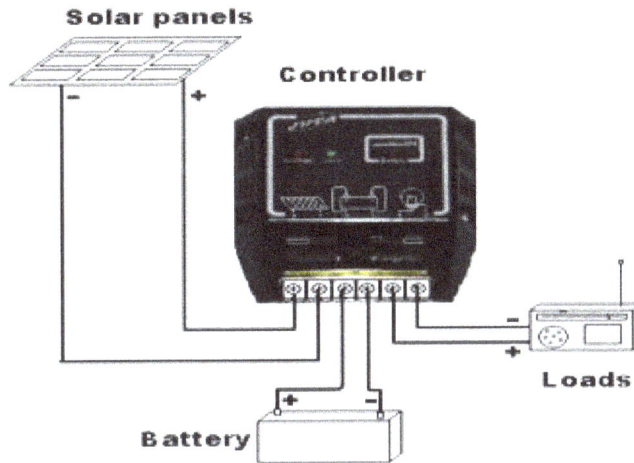

Solar charge controller

Applications

In recent days, the process of generating electricity from sunlight is having more popularity than other alternative sources and the photovoltaic panels are absolutely pollution free and they don't require high maintenance. The following are some examples where solar energy is utilizing.

- Street lights use photovoltaic cells to convert sunlight into DC electric charge. This system uses solar charge controller to store DC in the batteries and uses in many areas.

- Home systems use PV module for house-hold applications.

- Hybrid solar system uses for multiple energy sources for providing full time backup supply to other sources.

Example of Solar Charge Controller

From the below example, in this a solar panel is used to charge a battery. A set of operational amplifiers are used to monitor panel voltage and load current continuously. If the battery is fully charged, an indication will be provided by a green LED. To indicate under charging, over loading, and deep discharge condition a set of LEDs are used. A MOSFET is used as a power semiconductor switch by the solar charge controller to ensure the cut off load in low condition or over loading condition. The solar energy is bypassed using a transistor to a dummy load when the battery gets full charging. This will protect the battery from over charging.

This unit performs 4 major functions:

- Charges the battery,

- Gives an indication when battery is fully charged,

- Monitors the battery voltage and when it is minimum, cuts off the supply to the load switch to remove the load connection,

- In case of overload, the load switch is in off condition ensuring the load is cut off from the battery supply.

Hybrid Solar Charger

The efficiency of a solar charging system depends on the weather conditions. Solar panels generate the most electricity on clear days with abundant sunshine. Commonly, the solar panel gets four to five hours of bright sunlight in a day. If the weather is cloudy, it affects the battery charging process and the battery does not get a full charge.

This simple hybrid solar charger can give the solution for this problem. It can charge the battery using both solar power as well as AC mains supply. When the output from the solar panel is above 12 volts, the battery charges using the solar power and when the output drops below 12 volts, the battery charges through AC mains supply.

Hybrid Solar Charger Circuit

The below figure shows the hybrid solar charger circuit. The following hardware components required to build the hybrid solar charger circuit.

- A 12V, 10W solar panel (connected at SP1)

- Operational amplifier CA3130 (IC1)

- 12V single-changeover relay (RL1)

- 1N4007 Diodes

- Step-down transformer X1

- Transistor BC547 (T1)

- Few other RLC components

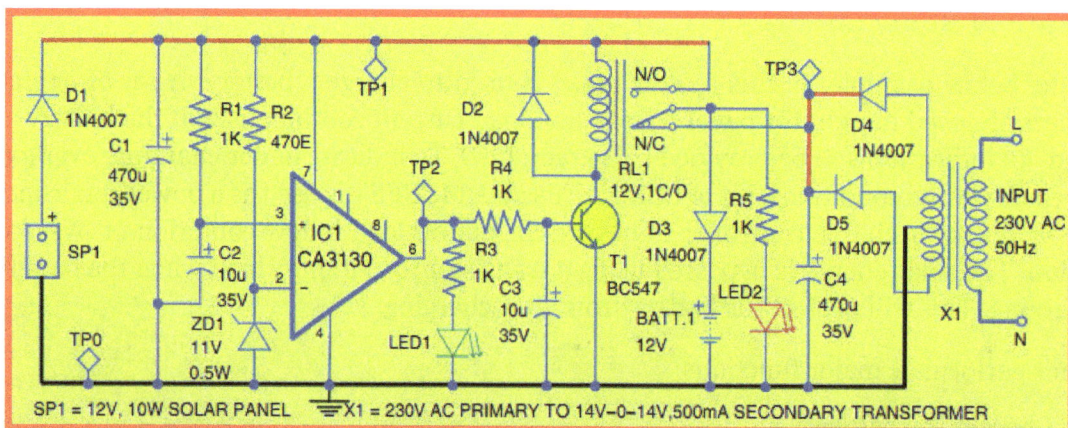

Hybrid Solar Charger Circuit

10 Watt, 12 Volt Solar Panel

In this circuit, we used a 10 Watt, 12 Volt Solar Panel. It will provide enough power to charge a 12V battery.

10 Watt, 12 Volt Solar Panel

This 10w-12v module is an array of 36 multi-crystalline silicon solar cells of similar performance, interconnected in series to obtain the 12-volt output.

These solar cells are mounted on a heavy duty anodized aluminium frame provides strength. For each 18 cells series strings, one bypass diode is installed. These cells are laminated between high transmissivity, low-iron, 3mm tempered glass and sheet of a Tedlar Polyester Tedlar (TPT) material by two sheets of ethylene Vinyl acetate (EVA). This setup protects against moisture penetrating into the module.

Key Features

- 36 high-efficiency Silicon Solar Cells
- Optimized module performance with Nominal Voltage 12 V DC
- Bypass diodes to avoid the hot spot effect
- Cells are embedded in a sheet of TPT and EVA
- Attractive, stable, heavy duty anodized aluminium frames with convenient
- Pre-cabled with fast-connecting systems

Hybrid Solar Charger Circuit Working

In sunny sunlight, the 12V, 10W solar panel delivers up to 17 volts DC with the 0.6-ampere current. The diode D1 provides reverse polarity protection and capacitor C1 buffers voltage from the solar panel. Op-amp IC1 is used as a simple voltage comparator.

Zener diode ZD1 provides a reference voltage of 11 volts to the inverting input of IC1. The non-inverting input of e op-amp gets voltage from the solar panel through R1.

The working of the circuit is simple. When the output from the solar panel is greater than or equal to 12 volts, Zener diode ZD1 conducts and provides 11 volts to the inverting terminal of IC1.

Since non-inverting input of the op-amp gets a higher voltage at this time, the output of the comparator turns high. Green LED1 glows when the comparator's output is high.

The transistor T1 then conducts and relay RL1 energized. Thus the battery gets charged current from the solar panel through the normally-open (N/O) and common contacts of relay RL1.

LED2 indicates charging of the battery. Capacitor C3 is provided for clean switching of transistor T1. Diode D2 protects transistor T1 from back EMF and diode D3 prevents the discharge of the battery current into the circuit.

When the output from the solar panel gets down below 12 volts, the output of the comparator turns low and the relay de-energizes. Now the battery gets charged current from the transformer based power supply through the normally closed (N/C) and common contacts of the relay.

This power supply includes step-down transformer X1, rectifying diodes D4 and D5, and smoothing capacitor C4.

Testing

To test the circuit for proper functioning, the below instructions to be followed:

- Remove the solar panel from connector SP1 and connect a DC variable voltage source.

- Set some voltage below 12V and slowly increase it.

- As the voltage reaches 12V and goes beyond, the logic at test point TP2 changes from low to high.

- The transformer-based power supply voltage can be checked at test point TP3.

Applications of Hybrid Solar Charger

In recent days, the process of generating electricity from sunlight has more popularity than other alternative sources and the photovoltaic panels are absolutely pollution free and they don't require high maintenance. The following are some examples.

- The Hybrid solar charger system used for multiple energy sources for providing full-time backup supply to other sources.

- Street lights use the solar cells to convert sunlight into DC electricity charge. This system uses a solar charge controller to store DC in the batteries and uses in many areas.

- Home systems use PV module for household applications.

Solar-powered Pump

A solar-powered pump is a pump running on electricity generated by photovoltaic panels or the radiated thermal energy available from collected sunlight as opposed to grid electricity or diesel run water pumps. The operation of solar powered pumps is more economical mainly due to the lower operation and maintenance costs and has less environmental impact than pumps powered by an internal combustion engine (ICE). Solar pumps are useful where grid electricity is unavailable and alternative sources (in particular wind) do not provide sufficient energy.

A windpump replaced by a solar powered pump at a water hole in the Augrabies Falls National Park

This solar water pump up to 5 hp is useful for farmers

Components

A photovoltaic solar powered pump system has three parts:

- Solar panels,
- Controller,
- Pump.

The solar panels make up most (up to 80%) of the systems cost. The size of the PV-system is directly dependent on the size of the pump, the amount of water that is required (m³/d) and the solar irradiance available.

The purpose of the controller is twofold. Firstly, it matches the output power that the pump receives with the input power available from the solar panels. Secondly, a controller usually provides a low voltage protection, whereby the system is switched off, if the voltage is too low or too high for the operating voltage range of the pump. This increases the lifetime of the pump thus reducing the need for maintenance. Other ancillary functions include automatically shutting down the system when water source level is low or when the storage tank is full, regulating water output pressure, blending power input between the solar panels and an alternate power source such as the grid or a petrol generator, and remotely monitoring and managing the system through an online portal offered as a cloud service by the manufacturer.

Voltage of the solar pump motors can be AC (alternating current) or DC (direct current). Direct current motors are used for small to medium applications up to about 4 kW rating, and are suitable for applications such as garden fountains, landscaping, drinking water for livestock, or small irrigation projects. Since DC systems tend to have overall higher efficiency levels than AC pumps of a similar size, the costs are reduced as smaller solar panels can be used.

Finally, if an alternating current solar pump is used, an inverter is necessary that changes the direct current from the solar panels into alternating current for the pump. The supported power range of inverters extends from 0.15 to 55 kW and can be used for larger irrigation systems. However, the panel and inverters must be sized accordingly to accommodate the inrush characteristic of an AC motor. To aid in proper sizing, leading manufacturers provide proprietary sizing software tested by third party certifying companies. The sizing software may include the projected monthly water output which varies due to seasonal change in insolation.

Water Pumping

Solar powered water pumps can deliver drinking water as well as water for livestock or irrigation purposes. Solar water pumps may be especially useful in small scale or community based irrigation, as large scale irrigation requires large volumes of water that in turn require a large solar PV array. As the water may only be required during some parts of the year, a large PV array would provide excess energy that is not necessarily required, thus making the system inefficient.

Solar PV water pumping systems are used for irrigation and drinking water. The majority of the pumps are fitted with a 2000 watt - 3,700 watt motor that receives energy from a 4,800 Wp PV array. The 5hp systems can deliver about 124,000 liters of water/day from a total of 50 meters setoff head and 70 meters dynamic head.

Oil and Gas

In order to combat negative publicity related to the environmental impacts of fossil fuels, including fracking, the industry is embracing solar powered pumping systems. Many oil and gas wells require the accurate injection (metering) of various chemicals under pressure to sustain their operation and to improve extraction rates. Historically, these chemical injection pumps (CIP) have been driven by gas reciprocating motors utilizing the pressure of the well's gas and exhausting the raw gas into the atmosphere. Solar powered electrical pumps (solar CIP) can reduce these greenhouse gas emissions. Solar arrays (photovoltaic cells) not only provide a sustainable power source for the CIPs but can also provide an electric source to run remote SCADA type diagnostics with remote control and satellite/cell communications from very remote locations to a desktop or notebook monitoring computer.

Stirling Engine

Instead of generating electricity to turn a motor, sunlight can be concentrated on the heat exchanger of a Stirling engine and used to drive a pump mechanically. This dispenses with the cost of solar panels and electric equipment. In some cases the Stirling engine may be suitable for local fabrication, eliminating the difficulty of importing equipment. One form of Stirling engine is the fluidyne engine which operates directly on the pumped fluid as a piston. Fluidyne solar pumps have been studied since 1987. At least one manufacturer has conducted tests with a Stirling solar powered pump.

Solar-powered Watch

A solar-powered watch/clock is a watch/clock that is powered by a solar cell, which converts light from the sun, or fluorescent lights, etc., into electrical energy. As shown in the figure below (wristwatch example), when the dial of the watch is exposed to light, the built-in solar cell generates electrical energy and stores it in the secondary cell (rechargeable battery) as a power source to run the watch. A solar-powered watch/clock is an ecological watch/clock that is friendly to people and the earth, harnessing non-polluting energy sources such as sunlight and using a rechargeable battery with the aim of reducing environmental burden.

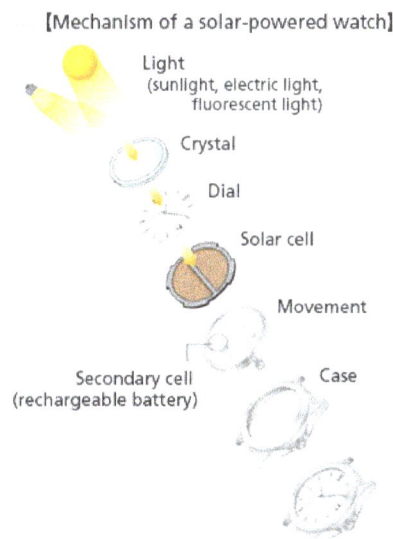

[Mechanism of a solar-powered watch]

Light
(sunlight, electric light,
fluorescent light)

Crystal

Dial

Solar cell

Movement

Secondary cell
(rechargeable battery)

Case

- Unlike a disposable battery (primary cell) such as dry battery and button battery, a secondary cell (rechargeable battery) is an earth-conscious electric cell that can be used over and over for a long period of time by recharging.

Solar watches are generally designed to be contemporary and usually have multiple functions. Solar watches are durable and many feature stainless steel or titanium cases and hardened mineral or sapphire crystals. Some feature a blue-grey tint on the dial, which is designed to optimize sunlight. Many solar watches emphasize their ruggedness by featuring a stopwatch, multiple time-zone displays and the ability to be water resistant up to 200 meters. Some solar watches feature a barometer, compass, thermometer and altimeter with LCD display. Solar-powered women's watches and men's watches offer functionality and style.

Photovoltaic Keyboard

A photovoltaic keyboard is a wireless computer keyboard that charges its batteries from a light source such as the sun or interior lighting, addressing a major drawback of wireless computer peripherals that otherwise require regular replacement of discharged batteries.

The Logitech Solar Keyboard Folio comes with a pre-installed integrated NiMH battery. On a full charge, the battery can last for up to 2500 hours of use. We recommend charging your tablet keyboard frequently to prevent the battery from running out while you're typing. Its wireless range Up to 30 feet or 10 meters. These are mostly in tablets. Wireless Solar Keyboard 750 makes battery and charging hassles a thing of the past. It takes advantage of any available light source to charge itself—and it's able to stay charged at least three months.

Logitech K750

The Logitech K750 has a set of photovoltaic cells on the top edge, charges in sunlight or under a standard bulb, can work up to three months in total darkness, and includes software to display battery charging status. It is a fullsized keyboard, including the usual movement keys and NumPad section on the right side, with low-profile keys much like a laptop. There are two models, compatible with Windows or Macintosh operating systems.

Even though the keyboard is not officially supported in Linux, a third party application named Solaar provides functionality akin to the original Logitech software, such as battery and connection status indications, and allows device pairing/unpairing.

Logitech K760

Another Logitech keyboard, the K760, is also PV powered, is highly useful. It is smaller and communicates with the computer via Bluetooth.

Advantage

- Self-charging wherever there's light,
- Charge lasts up to three months,
- Thin, space-saving profile,
- Key cap design makes for faster, quieter typing,

- Reliable, fast 2.4GHz wireless with 128-bit AES encryption,
- Unifying wireless receiver is so small you can leave it plugged in.

Solar-powered Calculator

The "Teal Photon", one of the first solar-powered calculators of the late 1970s (left) and a modern solar-powered scientific calculator (right)

Solar-powered calculators are hand-held electronic calculators powered by solar cells mounted on the device. They were introduced at the end of the 1970s.

Amorphous silicon has been used as a photovoltaic solar cell material for devices which require very little power, such as pocket calculators, because their lower performance compared to conventional crystalline silicon solar cells is more than offset by their lower cost and simplified deposition onto a substrate. The first solar powered calculators available in the late 1970s included the Royal *Solar 1*, the Sharp *EL-8026*, and the Teal *Photon*.

Solar calculators use liquid crystal displays as they are power efficient and capable of operating in the low voltage range of 1.5–2 V. Some models also use a light pipe to converge light onto the solar cells. However, solar calculators may not work well in indoor conditions under ambient lighting if sufficient light is not available.

Anylite Technology is the name of a solar technology used by Texas Instruments since the 1980s in some calculators. They are intended to be able to function with less light than other solar calculators. This was essentially achieved by using relatively large photovoltaic solar cells. The use of Anylite technology in modern TI calculators is denoted by a lower case "a" at the end of the model number (e.g. TI-30a). In older models, such as the TI-36 Solar, *Anylite Solar* is printed on the calculator.

As of the 2010s, some very cheap calculators include a "dummy" solar panel, implying that they are solar-powered, when they are actually powered only by battery.

Use in Solar Power Calculator

Nearly every basic calculator has a set of solar panels built into it. Even so, these calculators usually come with a battery inside that actually powers the device. These panels help extend the life of the calculator by slowly recharging the original battery. The manufacturer's intention is to make the calculator last long enough that it gets replaced for a reason other than the battery, often being dropped and broken.

Use most solar powered calculators like you would any other calculator; direct exposure to light typically doesn't matter.

Recharge a battery powered calculator by leaving it in a bright location but out of direct sunlight.

Recognize when you have a true solar calculator. The numbers will fade when the light is blocked and there will often be a noticeable lag between pushing a button and its appearance on a calculator. When using a true solar calculator operate it in a bright location but still out of direct sunlight.

Clean off the panels occasionally to keep them working well. Dust, dirt and oils from your fingers can cause a significant amount of refraction, and the solar energy is lost. Wipe them off with a paper towel or even the end of a shirt, the plastic covering over the panels will protect them.

Solar-powered Radio

A typical solar powered radio receiver

A solar powered radio is a portable radio receiver powered by photovoltaic panels. It is primarily used in remote areas where access to power sources is limited.

Advantages

Solar powered radios eliminate the need to replace batteries, which makes operating them cost much less. Since they don't require plugs, they can be used in areas where there is no electrical grid or generators. As a result, people in remote areas with little disposable income can have equal access to news and information. Informative radio programs on human rights, women's rights, the importance of education (especially for girls), HIV and AIDS, animal husbandry, agriculture, food security, combined with solar powered radios, can be a powerful tool for improving the lives of people in remote areas.

Solar-powered Fan

A solar fan is similar to the appliance store version of an attic or whole house fan, but has some amazing advantages that every homeowner should consider for their home.

The solar powered fan is basically a mechanical fan that uses the sun's energy to power the fan and remove hot air from your home. There are many benefits of installing a solar energy fan in your home, not the least of which is a more comfortable and healthier home.

Features and Specifications

A solar rechargeable fan is quite similar to a standard fan in terms of design. They are made up of lightweight materials, energy efficient motors, and have aerodynamically designed blades.

These fans are designed in two ways: table fans with a solar panel mounted directly on it or a separate panel that can be attached to it after being charged in the sunlight. The solar panels are primarily photovoltaic cells that absorb sunlight and convert it into electrical energy. The photovoltaic cells contain semiconductors like Silicon or Germanium which, after absorbing photons, emit electrons that in turn produce electricity. The electricity then gets transferred to the fan. The fan can not only be used during the day, but also during the night because it can store the solar power within the cells.

Uses of Solar Rechargeable Fan

Unlike conventional fans that merely circulate the air around, solar rechargeable fans have many different uses.

- Powerhouse ventilation

  Usually, a home with more than four members will have at least four fans, not counting the exhaust system. These consume a lot of electricity. Today, many people have installed solar panels on their rooftops that can power up not just their heaters but also the air circulation and ventilation system (solar rechargeable fans and exhausts).

- Outdoor use

  Solar rechargeable fans are great for outdoors and for those places that lack power outlets since they can run on the stored energy. Also, being portable, they can be placed according to your convenience.

- Travelling

  These fans are an excellent choice for travellers and adventurers who are always on the move. You can just charge up these fans during the day and use them anytime you please, no matter where you are. These fans are ideal for trekking and camping.

Different Types of Solar Rechargeable Fans

Like regular fans, solar rechargeable fans too have been diversified according to the function they would likely serve. Here are some varieties of solar rechargeable fans.

- Roof mounted attic fan

  Attic fans are meant to drive out excessive heat from the attic of a house and create an overall cooling effect throughout the house. These fans are attached to the roof of the house from where they draw the warm air and push it out of the house. Unlike electrical attic fans, solar roof mounted attic fans run on solar energy.

- Solar panel mounted fans

  These models of solar rechargeable fans have solar panels mounted on them. They cannot be detached from the fan body.

- Separate solar panel fans

  In these models, the solar panels are independent of the fan and can be attached to it with a wire. While the solar panel can be placed out under the sun, the fan can be kept inside for cooling.

Pedestal Solar Powered Fan

Pedestal fans are now available in solar powered versions. Growing line of solar powered stand fans offer a wide variety of sizes and features such as multi-speed airflow, left and right oscillation, and AC/DC automatic conversion.

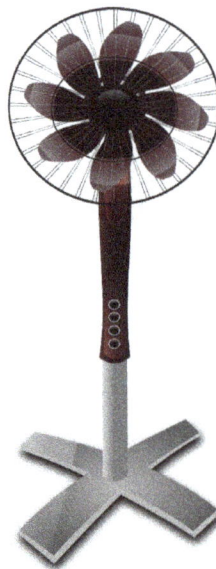

Floor Solar Fans

Our floor solar powered fans are an excellent tool for emergencies. The LED light bar and fan blades automatically turn on during a power failure, with the capability to run for up to 8 hours straight.

Desk Solar Power Fans

These medium-sized utility fans take on the same design cues as traditional desk fans but with mounted solar panels for storing power. They provide excellent air circulation with multiple air speeds and directionality. Some solar power fans even have additional USB ports for charging mobile phones and other electronic devices.

Solar Attic Fans

Solar attic fans are small fans installed on the roof of the house, and designed to remove the hot air that gathers in the attic of the home. These handy fans can be installed over your attic, in your garden shed, greenhouse or even your garage and they'll have the same effect – reducing heat and moisture in those spaces.

It's important to note that these solar fans are not meant to remove the hot air from within the house itself. Solar attic fans are also called solar exhaust fans, which work in the same way to remove hot air from an attic space.

Solar Whole House Fans

This fan is installed into the ceiling of the home and is designed to pull the hot air from inside the house and exhaust it out into the attic space.

In order to receive the most benefit from this type of fan, it should be operated during the early evening or overnight hours.

Solar Car Fans

Solar car fans are also referred to as solar vehicle ventilators. This style of solar fan uses sunlight to power a low-voltage fan, which is designed to push out the hot air that gathers inside a vehicle during the summer months.

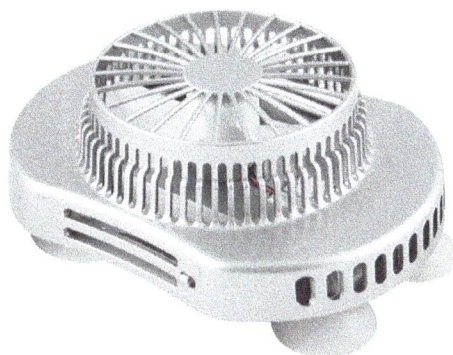

It is also supposed to draw fresh air from outside the vehicle, lowering the interior temperature by 15 degrees or so.

Some drawbacks with this type of solar powered fan: you must crack a window in order for it to work which could invite theft, they won't work in vehicles with tinted windows, and they don't seem to work on cloudy days.

Solar Gable Fans

Solar gable fans Work much like the solar attic fan except it is installed in the gable of a home.

Solar Hat Fans

These are also called solar fan hats or solar fan caps, are lots of fun to wear, and are just what they sound like.

The hat is outfitted with a mini sized solar panel, which is attached to the brim of the hat. The panel captures the energy of the sun and powers a small fan that is directed on your forehead and face. The hope is that the air it generates will cool you off, but the general consensus is that the benefit doesn't quite outweigh your appearance wearing the hat.

Benefits of Solar Operated Fans

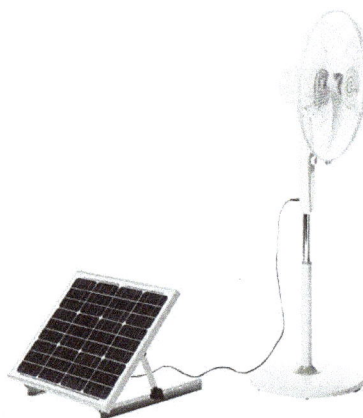

- Completely quiet when in operation,
- No electric connection required,
- Cuts down on the load your AC unit,

- Lower moisture levels and lower temperatures creates a healthier living space,

- Can adapt to almost any climate or environment,

- Reduces moisture and the risk of mold in your attic space,

- Can extend the life of your roof,

- There are tax incentives available in some areas of the country, which could lower your investment cost,

- Maintenance free,

- Costs nothing to operate once it's installed.

Drawbacks of Solar Operated Fans

As with any technology, there are a few drawbacks to solar operated fans:

- You may need to invest in more than one fan to pull enough hot air from the house to achieve a comfortable temperature,

- The sun needs to be shining in order for it to operate efficiently,

- Initial cost of solar version is higher than traditional one,

- Solar powered fans usually have a smaller motor and fan than electric version,

- Proper sealing is required or else air can be drawn from the house and moisture created inside the house,

- Homes that experience high winds aren't good candidates.

Transport Applications

Transport applications were amongst the first to benefit from the potential for photovoltaic generation and still account for a major share of world markets. The variety of applications and demands provides a microcosm of the whole photovoltaic market. Uses span principal transport media and include marine and radio beacons, railway crossings and transport communications systems.

Transport networks inevitably span large areas and cross regions with no established electricity grid, so benefit particularly from the flexibility of photovoltaic power sources. This has enabled new advances especially in marine and rail transport.

Railway authorities, notably in Australia and Canada have used the high reliability and low maintenance of photovoltaic systems to pioneer new applications in their field. Free power availability in remote regions naturally enables crossings to be protected with lights, audible signals and barriers automatically controlled from tract circuit sensors all powered by a solar array. Station-to-station and station-to-train communications may also be energized from sunshine.

Navigational aids, with their moderate intermittent power requirements are ideally suited to photovoltaic generation and such systems are already widely used. Experience in often hostile environments has contributed greatly to the supplier's knowledge in designing equipment which can operate reliably in high humid and saline conditions. Battery design and overall system integrity has proved as crucial as module performance. Meanwhile, the increasing competitiveness of photovoltaic devices has led to the viability of larger systems for radio beacons for example. Solar generators of a few kilowatts capacity are now seen to be economic for remote locations while moderate sizes are becoming viable even in inhabited areas.

Most transport systems demand high reliability as personal safety is often at stake. System design and overall integrity is thus the primary requirement. These factors are to be considered in detail. Users will demand and must receive exemplary engineering in these areas if the photovoltaic industry is to achieve its rightful place as a major energy supplier for transport applications.

Solar Vehicle

The solar vehicle is a step in saving non renewable sources of energy. The basic principle of solar vehicle is to use energy that is stored in a battery during and after charging it from a solar panel. The charged batteries are used to drive the motor which serves here as an engine and moves the vehicle in reverse or forward direction. The electrical tapping rheostat is provided so as to control the motor speed. This avoids excess flow of current when the vehicle is supposed to be stopped suddenly as it is in normal vehicle with regards to fuel.

Components used

Various types of electrical components were used for making the solar powered vehicle. A list of these components used with their range and the specific quantities that were required for making the solar vehicle is given in the following table.

Table: List of various components used

Components used	Range	Quantity
Batteries(heavy inverter batteries)	24V 190Ah	2*12V
Solar module	140Wp(Watt Peak)	1
Connecting Cables	Motor connection:-25Sq.m m high voltage cables.	10 meters
	Solar module to charge controller unit:-1Sq.mm	1 meter
	Charge controller to battery unit:-2Sq.mm	1 meter
Motor	High torque DC motor 1Hp=746W	1

Apart from the above listed components the main component that is responsible for speed control of the motor is the speed control switch. It is defined as follows:

• Speed control switch

The speed control of the DC motor is the essential part of the vehicle. For controlling speed of the motor, a switch was designed with 8 tapping, giving different values of resistance at each tapping, hence limiting the current that flows in the motor. The switch uses pure Nichrome wire for resistances. It uses a 8 tapping DC switch. The front view of the switch is as follows:-

Figure: The front view of the speed control unit

The switch has been provided with two terminals; one for the motor connections and the other for the battery connections. The arrangement of the switch is more or less like a rheostat. The different tapping act as resistance points. With each increase in the tapping value the value of resistance decrease, thus at the last tapping the motor will run at the highest speed as the limiting resistance will be minimum whereas the high torque condition of the motor will arise when the minimum tapping will be used, since the limiting resistance will be maximum.

The picture showing the view of the tapings is shown below in figure. It can be easily concluded that two coils are connected in a series to give one taping hence increasing the resistance.

Figure: The upper view if Speed control switch depicting the taping connections

The value of resistance at each taping is given in the table below. This resistance value is used for controlling the 1 hp motor.

Table: Values of resistance at various tapings of the switch

Terminal number	Resistance
1	435 milli-ohms
2	405 milli-ohms
3	358 milli-ohms
4	290 milli-ohms
5	220 milli-ohms
6	150 milli-ohms
7	076 milli-ohms
8	ZERO

Solar Panel – 140Wp

The solar panel used in the solar vehicle is of the rating of 140 WP. The main point that should be kept in mind while making a solar vehicle is the mounting of the solar panel. The panel should be mounted in such a way that it receives maximum sun rays so that it gives its maximum efficiency. For the vehicle designed, we have mounted the solar panel in SOUTH-EAST direction during the time 6 AM to 11.30 AM. After that the panel is changed to a SOUTH-WEST direction. We have used the conventional roof-top mounting technique for the solar panel A 6 feet by 4 feet plywood has been used and mounted on the top of vehicle.

The solar cell used in the vehicle is multi-crystalline. The reason behind using the multi crystalline cell is that it is more efficient than the mono-crystalline cell and the rate of conversion of energy is faster in the former. 36 cells are used in the PV module of this vehicle. The upper frame of this solar module is covered with thick glass to avoid breakage of the solar panel.

SOLAR MODULE- 140 Wp

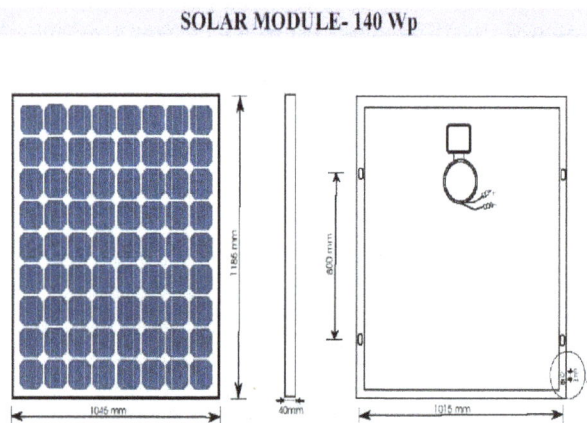

Figure: The diagrammatic representation of the panel and the panel connections

Working of the Vehicle

The solar module mounted on the top of car is used to charge the batteries via charge controller. A 140 WP solar module is used with output ranging from 24V to 25V at STC. The batteries are initially fully charged and then they are connected to solar module for charging. This helps to keep the battery charged always. This is also done as the efficiency of solar module is only 15%.Thus under this condition the battery gets fully charged again within 3hrs-3.5hrs. Thus to keep the full sine wave of charging this time lap is made. The maximum solar radiations are obtained between mornings 10am to evening 3:30pm. Hence, the panel is so mounted that maximum output may be obtained. As the supply is given through DPDT switch the motor takes a high starting current to propel the wheel to move in forward direction. On start the load on motor is nearly 250kg including the weight of person driving it. The motor after start acquires the maximum speed of 20kmph to 30kmph. The batteries get charged always from the solar panel and so it provides the continuous run for the vehicle. Motor must be started on top most gear so as to get maximum torque and speed to lift the full load. The speed may be varied later according to the driver's requirements. As the speed varies the load current also varies. So the speed variation must be low to keep battery alive for maximum duration of time. For stopping the motor, the speed control switch should be brought to minimum gear and then switch should be open; thereafter the mechanical brakes should be applied. The mechanical brakes can be applied instantly during emergency but this should be avoided as this could damage the motor and also produce unnecessary back emf. The average battery back-up is around four hours. The batteries are continuously charged by the solar panel but to increase their rate of charging three dynamos each of 24 V can be connected to the wheels of the vehicle. As the vehicle moves these dynamos will generate EMF and will charge the batteries. Hence the charging and discharging cycle of the batteries will be complete.

Land

Solar Cars

Solar cars depend on PV cells to convert sunlight into electricity to drive electric motors. Unlike solar thermal energy which converts solar energy to heat, PV cells directly convert sunlight into electricity.

The design of a solar car is severely limited by the amount of energy input into the car. Solar cars are built for solar car races and also for public use List of prototype solar-powered cars. Even the best solar cells can only collect limited power and energy over the area of a car's surface. This limits solar cars to ultra light composite bodies to save weight. Solar cars lack the safety and convenience features of conventional vehicles. The first solar family car was built in 2013 by students in the Netherlands. This vehicle is capable of 550 miles on one charge during sunlight. It weighs 850 pounds and has a 1.5kw solar array. Solar vehicles must be light and efficient. 3,000 pound or even 2,000 pound vehicles are less practical. Stella Lux, the predecessor to Stella, broke a record with a 932 mile single charge range. The Dutch are trying to commercialize this technology. During racing Stella Lux is capable of 700 miles during daylight. At 45 mph Stella Lux has infinite range. This is again due to high efficiency including a coefficient of drag of .16. The average family who never drive more than 200 miles a day would never need to charge from the mains. They would only plug in if they wanted to return energy to the grid. Solar cars are often fitted with gauges and/or wireless telemetry, to carefully monitor the car's energy consumption, solar energy capture and other parameters. Wireless telemetry is typically preferred as it frees the driver to concentrate on driving, which can be dangerous in such a small, lightweight car. The solar electric Vehicle system was designed and engineered as an easy to install (2 to 3 hours) integrated accessory system with a custom molded low profile solar module, supplemental battery pack and a proven charge controlling system.

As an alternative, a battery-powered electric vehicle may use a solar array to recharge; the array may be connected to the general electrical distribution grid.

Solar Buses

Solar buses are propulsed by solar energy, all or part of which is collected from stationary solar panel installations. The Tindo bus is a 100% solar bus that operates as free public transport service in Adelaide City as an initiative of the City Council. Bus services which use electric buses that are partially powered by solar panels installed on the bus roof, intended to reduce energy consumption and to prolong the life cycle of the rechargable battery of the electric bus, have been put in place in China.

Solar buses are to be distinguished from conventional buses in which electric functions of the bus such as lighting, heating or air-conditioning, but not the propulsion itself, are fed by solar energy. Such systems are more widespread as they allow bus companies to meet specific regulations, for example, the anti-idling laws that are in force in several of the US states, and can be retrofitted to existing vehicle batteries without changing the conventional engine.

Single-track Vehicles

The first solar "cars" were actually tricycles or Quadracycles built with bicycle technology. These were called solarmobiles at the first solar race, the Tour de Sol in Switzerland in 1985. With 72 participants, half used solar power exclusively while the other half used solar-human-powered hybrids. A few true solar bicycles were built, either with a large solar roof, a small rear panel, or a trailer with a solar panel. Later more practical solar bicycles were built with foldable panels to be set up only during parking. Even later the panels were left at home, feeding into the electric mains, and the bicycles charged from the mains. Today highly developed electric bicycles are available and these use so little power that it costs little to buy the equivalent amount of solar electricity. The "solar" has evolved from actual hardware to an indirect accounting system. The same system also works for electric motorcycles, which were also first developed for the Tour de Sol.

Applications

The Venturi Astrolab in 2006 was the world's first commercial electro-solar hybrid car, and was originally due to be released in January 2008.

In May 2007, a partnership of Canadian companies led by Hymotion altered a Toyota Prius to use solar cells to generate up to 240 watts of electrical power in full sunshine. This is reported as permitting up to 15 km extra range on a sunny summer day while using only the electric motors.

Nuna 3 PV powered car

An inventor from Michigan, USA built a street legal, licensed, insured, solar charged electric scooter in 2005. It had a top speed controlled at a bit over 30 mph, and used fold-out solar panels to charge the batteries while parked.

Auxiliary Power

Photovoltaic modules are used commercially as auxiliary power units on passenger cars in order to ventilate the car, reducing the temperature of the passenger compartment while it is parked in the sun. Vehicles such as the 2010 Prius, Aptera 2, Audi A8, and Mazda 929 have had solar sunroof options for ventilation purposes.

The area of photovoltaic modules required to power a car with conventional design is too large to be carried on board. A prototype car and trailer has been built Solar Taxi. According to the website, it is capable of 100 km/day using 6m² of standard crystalline silicon cells. Electricity is stored using a nickel/salt battery. A stationary system such as a rooftop solar panel, however, can be used to charge conventional electric vehicles.

It is also possible to use solar panels to extend the range of a hybrid or electric car, as incorporated in the Fisker Karma, available as an option on the Chevy Volt, on the hood and roof of "Destiny 2000" modifications of Pontiac Fieros, Italdesign Quaranta, Free Drive EV Solar Bug, and numerous other electric vehicles, both concept and production. In May 2007 a partnership of Canadian companies led by Hymotion added PV cells to a Toyota Prius to extend the range. SEV claims 20 miles per day from their combined 215W module mounted on the car roof and an additional 3kWh battery.

On 9 June 2008, the German and French Presidents announced a plan to offer a credit of 6-8g/km of CO_2 emissions for cars fitted with technologies "not yet taken into consideration during the standard measuring cycle of the emissions of a car". This has given rise to speculation that photovoltaic panels might be widely adopted on autos in the near future.

It is also technically possible to use photovoltaic technology, (specifically thermophotovoltaic (TPV) technology) to provide motive power for a car. Fuel is used to heat an emitter. The infrared radiation generated is converted to electricity by a low band gap PV cell (e.g. GaSb). A prototype TPV hybrid car was even built. The "Viking 29" was the World's first thermophotovoltaic (TPV) powered automobile, designed and built by the Vehicle Research Institute (VRI) at Western Washington University. Efficiency would need to be increased and cost decreased to make TPV competitive with fuel cells or internal combustion engines.

Personal Rapid Transit

JPods PRT concept with photovoltaic panels above guideways

Several personal rapid transit (PRT) concepts incorporate photovoltaic panels.

Rail

Railway presents a low rolling resistance option that would be beneficial of planned journeys and stops. PV panels were tested as APUs on Italian rolling stock under EU project.PVTRAIN. Direct feed to a DC grids avoids losses through DC to AC conversion. DC grids are only to be found in electric powered transport: railways, trams and trolleybuses. Conversion of DC from PV panels to grid alternating current (AC) was estimated to cause around 3% of the electricity being wasted.

PVTrain concluded that the most interest for PV in rail transport was on freight cars where on board electrical power would allow new functionality:

- GPS or other positioning devices, so as to improve its use in fleet management and efficiency.

- Electric locks, a video monitor and remote control system for cars with sliding doors, so as to reduce the risk of robbery for valuable goods.

- ABS brakes, which would raise the maximum velocity of freight cars to 160 km/h, improving productivity.

The Kismaros – Királyrét narrow-gauge line near Budapest has built a solar powered railcar called 'Vili'. With a maximum speed of 25 km/h, 'Vili' is driven by two 7 kW motors capable of regenerative braking and powered by 9.9m2 of PV panels. Electricity is stored in on-board batteries. In addition to on-board solar panels, there is the possibility to use stationary (off-board) panels to generate electricity specifically for use in transport.

A few pilot projects have also been built in the framework of the "Heliotram" project, such as the tram depots in Hannover Leinhausen and Geneva (Bachet de Pesay). The 150 kW$_p$ Geneva site injected 600V DC directly into the tram/trolleybus electricity network provided about 1% of the electricity used by the Geneva transport network at its opening in 1999. On December 16, 2017 a fully solar-powered train was launched in New South Wales, Australia. The train is powered using onboard solar panels and onboard rechargeable batteries. It holds a capacity for 100 seated passenger for a 3 km journey.

Recently Imperial College London and the environmental charity 10:10 have announced the Renewable Traction Power project to investigate using track-side solar panels to power trains . Meanwhile, Indian railways announced their intention to use on board PV to run air conditioning systems in railway coaches. Also, Indian Railways announced it is to conduct a trial run by the end of May 2016. It hopes that an average of 90,800 liters of diesel per train will be saved on an annual basis, which in turn results in reduction of 239 tones of CO_2.

Water

Planet Solar, the world's largest solar-powered boat and the first ever solar electric vehicle to circumnavigate the globe.

Solar powered boats have mainly been limited to rivers and canals, but in 2007 an experimental 14m catamaran, the Sun21 sailed the Atlantic from Seville to Miami, and from there to New York. It was the first crossing of the Atlantic powered only by solar.

Japan's biggest shipping line Nippon Yusen KK and Nippon Oil Corporation said solar panels capable of generating 40 kilowatts of electricity would be placed on top of a 60,213 ton car carrier ship to be used by Toyota Motor Corporation.

In 2010, the Tûranor Planet Solar, a 30 metre long, 15.2 metre wide catamaran yacht powered by 470 square metres of solar panels, was unveiled. It is, so far, the largest solar-powered boat ever built. In 2012, PlanetSolar became the first ever solar electric vehicle to circumnavigate the globe.

Various demonstration systems have been made. Curiously, none yet takes advantage of the huge power gain that water cooling would bring.

The low power density of current solar panels limits the use of solar propelled vessels, however boats that use sails (which do not generate electricity unlike combustion engines) rely on battery power for electrical appliances (such as refrigeration, lighting and communications). Here solar panels have become popular for recharging batteries as they do not create noise, require fuel and often can be seamlessly added to existing deck space.

Air

There is considerable military interest in unmanned aerial vehicles (UAVs); solar power would enable these to stay aloft for months, becoming a much cheaper means of doing some tasks done today by satellites. In September 2007, the first successful flight for 48h under constant power of a UAV was reported. This is likely to be the first commercial use for photovoltaics in flight.

Swiss solar-powered aircraft Solar Impulse completed a circumnavigation of the world in 2016.

Gossamer Penguin

Solar ships can refer to solar powered airships or hybrid airships.

Many demonstration solar aircraft have been built, some of the best known by AeroVironment.

Manned Solar Aircraft

- Gossamer Penguin,

- Solar Challenger: This aircraft flew 163 miles (262 km) from Paris, France to England on solar power.

- Sunseeker

- Solar Impulse: Two single-seat aircraft, the second of which circumnavigated the Earth. The first aircraft completed a 26-hour test flight in Switzerland on 8-9 July 2010. The aircraft was flown to a height of nearly 28,000 feet (8,500 meters) by Andre Borschberg. It flew overnight using battery power. The second aircraft, slightly larger and more powerful, took off from Abu Dhabi in 2015, flew towards India and then eastward across Asia. However, after experiencing battery overheating, it was forced to halt in Hawaii over the winter. In April 2016, it resumed its journey, and completed its circumnavigation of the globe, returning to Abu Dhabi on 26 July 2016.

- SolarStratos: Swiss stratospheric 2-seater solar plane aims to climb into space.

Hybrid Airships

An Australian-based company is working on a project to develop an air crane called the SkyLifter, a "vertical pick-up and delivery aircraft" being capable of lifting up to 150 tons.

A Canadian start-up, Solar Ship Inc, is developing solar powered hybrid airships that can run on solar power alone. The idea is to create a viable platform that can travel anywhere in the world delivering cold medical supplies and other necessitates to locations in Africa and Northern Canada without needing any kind of fuel or infrastructure. The hope is that technology developments in solar cells and the large surface area provided by the hybrid airship are enough to make a practical solar powered aircraft. Some key features of the Solarship are that it can fly on aerodynamic lift alone without any lifting gas, and the solar cells along with the large volume of the envelope allow the hybrid airship to be reconfigured into a mobile shelter that can recharge batteries and other equipment.

The Hunt *GravityPlane* is a proposed gravity-powered glider by Hunt Aviation in the USA. It also has aerofoil wings, improving its lift-drag ratio and making it more efficient. The GravityPlane requires a large size in order to obtain a large enough volume-to-weight ratio to support this wing structure, and no example has yet been built. Unlike a powered glider, the GravityPlane does not consume power during the climbing phase of flight. It does however consume power at the points where it changes its buoyancy between positive and negative values. Hunt claim that this can nevertheless improve the energy efficiency of the craft, similar to the improved energy efficiency of underwater gliders over conventional methods of propulsion. Hunt suggest that the low power consumption should allow the craft to harvest sufficient energy to stay aloft indefinitely. The conventional approach to this requirement is the use of solar panels in a solar-powered aircraft. Hunt has proposed two alternative approaches. One is to use a wind turbine and harvest energy from the airflow generated by the gliding motion, the other is a thermal cycle to extract energy from the differences in air temperature at different altitudes.

Unmanned Aerial Vehicles

- Pathfinder and Pathfinder-Plus - This UAV demonstrated that an airplane could stay aloft for an extended period of time fueled purely by solar power.

- Helios: Derived from the Pathfinder-Plus, this solar cell and fuel cell powered UAV set a world record for flight at 96,863 feet (29,524 m).

- Zephyr: Built by Qinetiq, this UAV set the unofficial world record for longest duration unmanned flight at over 82 hours on 31 July 2008. Just 15 days after the Solar Impulse flight mentioned above, on 23 July 2010 the Zephyr, a lightweight unmanned aerial vehicle engineered by the United Kingdom defence firm QinetiQ, claimed the endurance record for an unmanned aerial vehicle. It flew in the skies of Arizona for over two weeks (336 hours). It has also soared to over 70,700 feet (21.5 km).

- China's designed and manufactured UAV successfully reached an altitude of 20,000 meters during a test flight in the country's northwest regions. Named "Caihong" (CH), or "Rainbow" in English, it was developed by a research team from CASC.

Future Projects

- The Persistent High Altitude Solar Aircraft Phasa-35 being developed by BAE Systems & aerospace technology firm Prismatic for test flights in 2019.

- Titan Aerospace acquired by Google aimed to develop the Solar UAV, however the project seems to be abandoned

- Sky-Sailor (aimed at Martian flight)

- Various solar airship projects, such as Lockheed Martin's "High Altitude Airship"

Space

Solar Powered Spacecraft

PV on the International Space Station

Solar energy is often used to supply power for satellites and spacecraft operating in the inner solar system since it can supply energy for a long time without excess fuel mass. A Communications satellite contains multiple radio transmitters which operate continually during its life. It would be uneconomic to operate such a vehicle (which may be on-orbit for years) from primary batteries or fuel cells, and refuelling in orbit is not practical. Solar power is not generally used to adjust the satellite's position, however, and the useful life of a communications satellite will be limited by the on-board station-keeping fuel supply.

Solar Propelled Spacecraft

A few spacecraft operating within the orbit of Mars have used solar power as an energy source for their propulsion system.

All current solar powered spacecraft use solar panels in conjunction with electric propulsion, typically ion drives as this gives a very high exhaust velocity, and reduces the propellant over that of a rocket by more than a factor of ten. Since propellant is usually the biggest mass on many spacecraft, this reduces launch costs.

Other proposals for solar spacecraft include solar thermal heating of propellant, typically hydrogen or sometimes water is proposed. An electrodynamic tether can be used to change a satellite's orientation or adjust its orbit.

Another concept for solar propulsion in space is the light sail; this doesn't require conversion of light to electrical energy, instead relying directly on the tiny but persistent radiation pressure of light.

Planetary Exploration

Perhaps the most successful solar-propelled vehicles have been the "rovers" used to explore surfaces of the Moon and Mars. The 1977 Lunokhod programme and the 1997 Mars Pathfinder used solar power to propel remote controlled vehicles. The operating life of these rovers far exceeded the limits of endurance that would have been imposed, had they been operated on conventional fuels.

Electric Vehicle with Solar Assist

Louis Palmer standing in the Solartaxi.

A Swiss project, called "Solartaxi", has circumnavigated the world. This is the first time in history an electric vehicle (not self sufficient solar vehicle) has gone around the world, covering 50000 km in 18 months and crossing 40 countries. It is a road-worthy electric vehicle hauling a trailer with solar panels, carrying a 6 m² sized solar array. The Solartaxi has Zebra batteries, which permit a range of 400 km without recharging. The car can also run for 200 km without the trailer. Its maximum speed is 90 km/h. The car weighs 500 kg and the trailer weighs 200 kg. According to initiator and tour director Louis Palmer, the car in mass production could be produced for 16000 Euro. Solartaxi has toured the World from July 2007 till December 2008 to show that solutions to stop global warming are available and to encourage people in pursuing alternatives to fossil fuel. Palmer suggests the most economical location for solar panels for an electric car is on building rooftops though, likening it to putting money into a bank in one location and withdrawing it in another.

Solar electrical vehicles is adding convex solar cells to the roof of hybrid electric vehicles.

Plug-In Hybrid and Solar Vehicles

An interesting variant of the electric vehicle is the triple hybrid vehicle—PHEV that has solar panels as well to assist.

The 2010 Toyota Prius model has an option to mount solar panels on the roof. They power a ventilation system while parked to help provide cooling. There are many applications of photovoltaics in transport either for motive power or as auxiliary power units, particularly where fuel, maintenance, emissions or noise requirements preclude internal combustion engines or fuel cells. Due to the limited area available on each vehicle either speed or range or both are limited when used for motive power.

PV used for auxiliary power on a yacht

Limitations

There are limits to using photovoltaic (PV) cells for vehicles:

- Power density: Power from a solar array is limited by the size of the vehicle and area that can be exposed to sunlight. This can also be overcome by adding a flatbed and connecting it to the car and this gives more area for panels for powering the car. While energy can be accumulated in batteries to lower peak demand on the array and provide operation in sunless conditions, the battery adds weight and cost to the vehicle. The power limit can be mitigated by use of conventional electric cars supplied by solar (or other) power, recharging from the electrical grid.

- Cost: While sunlight is free, the creation of PV cells to capture that sunlight is expensive. Costs for solar panels are steadily declining (22% cost reduction per doubling of production volume).

- Design considerations: Even though sunlight has no lifespan, PV cells do. The lifetime of a solar module is approximately 30 years. Standard photovoltaics often come with a warranty of 90% (from nominal power) after 10 years and 80% after 25 years. Mobile applications are unlikely to require lifetimes as long as building integrated PV and solar parks. Current PV panels are mostly designed for stationary installations. However, to be successful in mobile applications, PV panels need to be designed to withstand vibrations. Also, solar panels, especially those incorporating glass, have significant weight. In order for its addition to be of value, a solar panel must provide energy equivalent to or greater than the energy consumed to propel its weight.

Advantages of the Vehicle

The solar vehicles are the future of the automobile industry. They are highly feasible and can be manufactured with ease. The main advantages of a solar vehicle are that they are pollution less and are very economical. Since they cause no pollution they are very eco-friendly and are the only answer to the increasing pollution levels from automobiles in the present scenario. By harvesting the renewable sources of energy like the solar energy we are helping in preserving the non-renewable sources of energy. The other main advantages of the solar vehicle are that they require less maintenance as compared to the conventional automotive and are very user friendly.

References

- McDermott, James E. Horne; Maura (2001). The next green revolution : essential steps to a healthy, sustainable agriculture. New York [u.a.]: Food Products Press. p. 226. ISBN 1560228865.

- Photovoltaics-solar-panel-uses-and-application-examples: apva.org.au, Retrieved 25 June 2018

- Kudo, Yuya; Shonchoy, Abu S.; Takahashi, Kazushi. "Can Solar Lanterns Improve Youth Academic Performance? Experimental Evidence from Bangladesh". The World Bank Economic Review. doi:10.1093/wber/lhw073.

- History-of-solar-powered-refrigerator: historyofrefrigeration.com, Retrieved 15 July 2018

- Wethe, David (29 November 2012). "For Fracking, It's Getting Easier Being Green". Bloomberg Businessweek. Retrieved 12 December 2012.

- Solar-refrigeration-13698: coolingindia.in, Retrieved 11 April 2018

- "Solar-powered rail vehicle ready for service". International Railway Journal. May 20, 2013. Retrieved 2013-05-20.

- What-is-solar-air-conditioning: canstarblue.com.au, Retrieved 31 March 2018

- Goetzberger, Adolf (1998). Crystalline silicon solar cells. Germany: John Wiley&Sons. p. 72. ISBN 0-471-97144-8.

- Solar-charge-controller: elprocus.com, Retrieved 25 June 2018

- "The potential of solar powered transportation and the case for solar powered railway in Pakistan". Renewable and Sustainable Energy Reviews. Elsevier. 39: 270–276. doi:10.1016/j.rser.2014.07.025. Retrieved 2013-05-22.

- Faqs-about-solar-watches: overstock.com, Retrieved 15 July 2018

- "All-Electric, Solar-Powered, Free Bus!!!". Ecogeek.org. 2007-12-27. Archived from the original on 2009-09-08. Retrieved 2013-01-12.

- Solar-powered-fan: solarlightsmanufacturer.com, Retrieved 19 March 2018

- Bergesen, Joseph D.; Tähkämö, Leena; Gibon, Thomas; Suh, Sangwon (2016). "Potential Long-Term Global Environmental Implications of Efficient Light-Source Technologies". Journal of Industrial Ecology. 20 (2): 263. doi:10.1111/jiec.12342.

Permissions

All chapters in this book are published with permission under the Creative Commons Attribution Share Alike License or equivalent. Every chapter published in this book has been scrutinized by our experts. Their significance has been extensively debated. The topics covered herein carry significant information for a comprehensive understanding. They may even be implemented as practical applications or may be referred to as a beginning point for further studies.

We would like to thank the editorial team for lending their expertise to make the book truly unique. They have played a crucial role in the development of this book. Without their invaluable contributions this book wouldn't have been possible. They have made vital efforts to compile up to date information on the varied aspects of this subject to make this book a valuable addition to the collection of many professionals and students.

This book was conceptualized with the vision of imparting up-to-date and integrated information in this field. To ensure the same, a matchless editorial board was set up. Every individual on the board went through rigorous rounds of assessment to prove their worth. After which they invested a large part of their time researching and compiling the most relevant data for our readers.

The editorial board has been involved in producing this book since its inception. They have spent rigorous hours researching and exploring the diverse topics which have resulted in the successful publishing of this book. They have passed on their knowledge of decades through this book. To expedite this challenging task, the publisher supported the team at every step. A small team of assistant editors was also appointed to further simplify the editing procedure and attain best results for the readers.

Apart from the editorial board, the designing team has also invested a significant amount of their time in understanding the subject and creating the most relevant covers. They scrutinized every image to scout for the most suitable representation of the subject and create an appropriate cover for the book.

The publishing team has been an ardent support to the editorial, designing and production team. Their endless efforts to recruit the best for this project, has resulted in the accomplishment of this book. They are a veteran in the field of academics and their pool of knowledge is as vast as their experience in printing. Their expertise and guidance has proved useful at every step. Their uncompromising quality standards have made this book an exceptional effort. Their encouragement from time to time has been an inspiration for everyone.

The publisher and the editorial board hope that this book will prove to be a valuable piece of knowledge for students, practitioners and scholars across the globe.

Index